The narrative is based on real historical figures; however, the dialogue and depiction of interpersonal relationships are fictional.

The facts throughout this publication are attributed to the original works in the references section. All references were in the public domain and did not include a nonuse statement at the time of publication.

The science depicted in this publication is accurate to current theories.

First Edition

ISBN : 979-8-9988536-0-9 Hardback
ISBN : 979-8-9988536-1-6 Paperback

History, Repeating Itself

By Colby Kerstetter

"The privilege of a lifetime is to become
who you truly are."

-Dr. Carl Gustav Jung

Table of Contents

कर्म Karma

Language of Origin: Sanskrit

Definition: action, doing

Buddhist interpretation: an intention [a goal] which is acted upon,
that leads to future consequences or
rewards in saṃsāra

संसार Saṃsāra

Language of Origin: Sanskrit

Definition: world

Buddhist interpretation: our reality as experienced by each soul, in a
constant cycle of reincarnation, in which,
we are destined to the karma built by our
previous lives

Prelude

Karma can be a tricky thing. It's the old mantra we all know; what goes around comes around, but it has another meaning. Karma lays out a path for our lives that we are drawn to follow. Our free will allows us to deviate from this path, but our karma pulls us to it. The funny thing is, when we make choices that keep us on our karmic path, we call it destiny.

Sometimes karma challenges us with the same trial as our previous life, like a test we must pass to get one step closer to enlightenment, whatever that is. This repetitive cycle of trials can be like playing a video game. You are trying to beat a certain level, but you keep dying at the same spot until you make a change. In life you are presented with the same challenge until you finally make that change and fulfill your purpose.

It isn't just individuals who ignorantly repeat their errors. Entire societies have repeated the same devastating mistakes as previous ones, but today most of us are not concerned with obscure memories of long-lost civilizations. After all, we are much too advanced to repeat history's mistakes; aren't we?

This karmic repetition through time is usually subtle but even when it's blatant it hides in plain sight. Just ask Wolfgang Paul who won the Nobel Prize in Physics in 1989. Not to be confused with Wolfgang Pauli who won the Nobel Prize in Physics in 1945. [FACT 1]

Are the Wolfgangs too subtle for your liking? The curse of the Washington Redskins will prove otherwise. On November 18th their quarterback was sacked, breaking his right tibia and fibula. It happened around the 40-yard line at their home field, caused by a player who had been named Defensive-Player-of-the-Year three times, while the quarterback's left tackle was off the field due to injury. The Redskins lost the game with a score of 23 to 21.

You may recognize this as Joe Theismann's career ending injury and you would be correct, but history repeated itself. This entire scenario played out again 33 years later exactly as it did the first time. On November 18th Redskins quarterback, Alex Smith, was sacked, breaking his right tibia and fibula. It happened around the 40-yard line at their home field, caused by a player who had been named Defensive-Player-of-the-Year three times, while the quarterback's left tackle was off the field due to injury. The Redskins lost the game with a score of 23 to 21. [FACT 2]

Within these pages you will realize mere coincidence and random chance are intertwined with the very fabric of reality, only there is nothing mere about them.

The whole universe, and you within it, are the product of billions of years of evolution, which gives you an unknowingly personal connection to history. From the fall of Rome to a single butterfly fluttering its wings, each action is like a ripple on a pond. When all those ripples overlap, some cancel out while others combine, creating bigger ripples, eventually becoming waves. All your experiences today have been shaped by those intersecting waves that intensified from nothing.

Whether acknowledged or not, we are each a part of history, repeating itself.

History, Repeating Itself starts with Part I, *The Owl and the Osprey*, which bridges religious spirituality with the latest scientific theories so you can form a deep understanding of the universe and your part in it.

You will find that not all characters have both a name and a speaking role. Some characters speak without being named and others are named but never speak. Characters who speak and are named will be reincarnated in Part II (*Treasure Lies Within*) and Part III (*Into the Wild We Go*). Follow along through these individuals' incarnations and see how history repeats itself.

FACTS

1.	In 1945 Wolfgang Pauli won the Nobel Prize in physics after his good friend, Albert Einstein, nominated him for discovering that electrons cannot simultaneously occupy the same quantum state. In truth, the Pauli Exclusion Principle is more complicated than that, but this mathematical realization paved the way for a deeper understanding of the physics that govern our reality. [Ref 1]

In 1989 Wolfgang Paul shared the Nobel Prize in Physics for his work with the Paul trap, which isolated particles so we can measure them more accurately. [Ref 1]

Paul met Pauli at a physics conference in the 1950's where he joked that he was Pauli's imaginary part. The joke was inspired by the difference in their names. In math an "i" represents the mathematical function 'imaginary'. [Ref. 1]

2.   This was a coincidence 33 years in the making. In 1985 and 2018, both on November 18th, the quarterback on the Washington Redskins, now the Commanders, was injured in the most peculiarly similar way. [Ref 2]

The coincidences are listed here:

-The games were on November 18th

-Quarterback injured by a sack

-Broke right tibia and fibula

-Around the 40-yard line at their home field

-Sacked by a three time Defensive-Player-of-the-Year

-The quarterback's left tackle was off the field due to injury

-The Redskins lost the game 23 to 21 in both games

Part I

The Owl and the Osprey

Based on true stories of real people.

Forward

I am here to tell you there is life after death. I cannot say if there is a heaven or hell but what I do know is the thing we call consciousness is part of something much larger than ourselves that exists beyond our physical bodies and transcends our individual perceptions.

Many scientists claim there is no reason to suspect consciousness is anything other than our everyday experiences but in this book, you will see real examples of peculiar things that science cannot yet explain; coincidences too fitting to be random chance, a knife that shattered as it sat idle on the kitchen counter, and streetlights that seem to turn off when certain people walk by.

No science or math has ever disproven the possibility of life after death. Some claim consciousness does not exist outside the physical body because there is no evidence. An absence of evidence is not evidence of absence, but in truth, there is evidence. Have you been paying attention to the peculiar things happening around you? They might be more than ordinary coincidence.

Enjoy.

Chapter 1

Impending Doom

"I profess no belief. I know there are experiences that
one must pay religious attention to."

-Dr. Carl G. Jung
Letters Vol II, Page 517

1881, Switzerland

It was a Sunday in the middle of summer as a young boy named Carl
sat in the front pew of his church, trying not to doze off. The modest chapel
was the center of a peaceful Swiss village, nestled in the hills along the Rhine
River. The church stifled in the sun until someone in the back quietly
propped open the old wooden doors. A cool breeze rushed in and chased the
heat through the open windows in the front. The preacher's voice droned on
and as the preacher's son, Carl was all too familiar with the sermon.

As he sat in the pew with his mother, the cool breeze blew gently
over his face, until he could no longer fight the urge to drift away. His
father's monotone voice faded, and six-year-old Carl imagined he was playing
on the riverbank at the edge of town.

The water gurgled in shallow pools along the river's edge around him.
Other children laughed as they threw stones, seeing who could make the
biggest splash. Behind him, the great Alps jutted into the air as if they stood
in defiance of the sky itself. Across the river, in Germany, cows grazed in the
meadow of tall grass and wild summer flowers that grew on the side of a hill.
Large brass bells hung around their necks clanging chaotically like a group of
drummers, each slowly playing their own song.

Carl noticed a rather large yellow flower swaying in the summer breeze until a cow chomped it from its stem along with a mouthful of grass. In an instant the summer flowers and the green pastures disappeared. The boy awoke as his mother's hand hit the back of his blond head.

"What would the congregation think of your father if his own son sleeps through the sermon?" She whispered angrily as his father's preaching finally ended.

Carl had a nice childhood in the sleepy village but at home his father was worried about much more sinister things than his son sleeping in the front pew. Late one morning, Carl entered the small kitchen looking for some lunch. He stopped dead in his tracks as he heard his father scolding his mother.

"The neighbor told me you were outside last night yelling gibberish at him! Why? What has he ever done to you?"

This wasn't the first time this sort of thing happened. Little Carl was used to it by now. His mother seemed normal, but she had a strange air about her. Odd things would happen to her or maybe even because of her.

"I remember being out back when a man appeared from the side of the house." His mother replied weakly. "He had an awful headache. He didn't say anything, he just grabbed his head like he was in agony, and I knew."

"There was no one else!" His father roared. "The neighbor didn't have a headache and when I found you, you were staring at the back of our house muttering nonsense to yourself! What was going through your head?" He demanded without waiting for an answer. "And what's worse is Carl saw you. I don't want you teaching him this nonsense!"

She sounded defeated, "I can't help it. I see strange things at night. Some evenings I walk in the house and the next thing I know I'm waking up the next morning. I don't know why I can't remember." She began to sob.

"Well from now on you need to come inside before nightfall and stay there! I am the pastor! My profession relies on this family's reputation! Another outburst like that and we will have to sell everything and move... AGAIN!"

4

Impending Doom

The summer pressed on and Carl forgot all about his parent's fighting. One evening after dinner, he heard the other children playing outside and went to join them. In Switzerland, twilight lingers in the summertime, almost unnaturally long like one of Carl's father's sermons.

The kids danced and played in the setting sun, squeezing the day for every last drop of light before nightfall. When Carl got back home the orange sunlight was fading to a pale yellow which lingered on the horizon like the glowing embers in a burned-out fire.

Inside was considerably darker than outside. They had no electric lights to turn on and with his father leading bible study at church, no one had lit any of the gas lamps. There was just enough twilight creeping through the windows for Carl to see the hallway and the stairs. It was the exact time of evening when your eyes start to play tricks on you. Carl began his ascent up the staircase towards his bedroom but there was an odd feeling in the house. He walked slowly while his eyes adjusted. Every step of the winding staircase let out a hissing creak as if the house was screeching for help.

He rounded the stairs, stopping on the second floor between a window and the hall to his parent's room. He heard an odd noise humming down the otherwise empty corridor. Carl froze. He recognized the sound. It was like a mix between humming and growling and there was no doubt, his mother was in a trance.

For a moment he stood there staring into the emptiness of the hallway, paralyzed by an anxious curiosity. Thinking back on it later, he was not sure why he stood there listening to the humming. He should have continued to his room further up the staircase, but it was like something wanted him to stay. What he saw next, he would never forget.

A large shadowy figure with broad shoulders emerged from his parents' bedroom. It didn't seem to be walking, instead it just floated through the doorway. The figure stopped and faced Carl. A hot rush of adrenaline shot down Carl's spine and coursed through his veins. His hair stood on end as goosebumps spread up his neck and down his arms. There was electricity in the air.

It resembled the shape of a person but somehow it seemed out of focus, almost as if it didn't fully exist.

Who is this? Carl thought, staring directly at the shadow's face, in defiance of his fear. Instead of a face, the shadowy figure had a blue glow without any definition or features. Carl was enthralled with this glowing specter whose light danced on the hallway walls.

The pale blue orb of light grew larger, fully encircling the being's shadowy head, then it happened. The thing's head lifted clean off its body and rose towards the ceiling. The light surrounding the head was slightly faded as if Carl was looking at a full moon rising in the night's sky obscured by thick puffy clouds. The headless figure stood frozen in space and time. The dark shadowy circle of a head shrunk smaller and smaller.

Carl was not scared, something within him understood there was no danger. After all, this was not the first time something strange and unexplainable happened around his mother during the night. As the floating head shrank, he looked back to the headless body where a new glowing head was sprouting upon the shoulders of this odd figure. Once the head was full size it too floated off the body and shrank into nothingness.

Carl started to count and no less than six times he saw a head grow upon the shoulders of the apparition before floating into the air and dissolving into oblivion. [FACT 3]

"It is a cycle." Carl whispered aloud. "He is trapped in a cycle."

This ghost or spirit, or perhaps it was some sort of disembodied persona, but whatever it was, Carl understood it was somehow an extension of his mother. He felt like it was trapped, a victim, perhaps of its own actions. Carl was overcome with the feeling that he needed to help break the cycle of what this thing was experiencing. He took a step forward, stomping his foot on the wooden planks of the hallway like a bull about to charge. The shadowy figure instantly disappeared without a trace, as if it had never been there at all.

Carl was left stunned, standing in total darkness. The phantom's blue glow vanished and with it, Carl realized the hazy yellow light from the sunset was gone too. When he first saw the thing, it felt like time stopped but now he realized the sun had set and it felt like time caught up with itself all at once. Was it all in my head, he wondered. Were my eyes playing tricks on me?

He realized the humming noise from inside the bedroom had also stopped and a peaceful silence fell throughout the house. Before laying down Carl opened his bedroom window and listened to the crickets singing and the babbling Rhine River flowing through the valley below.

"Your mother is not well." Carl's father said the next morning. "She told me what happened last night. She will be going to a hospital in Basel where she will get the help she needs." [FACT 4]

At the time, Carl was only six years old. He didn't understand how or why, but he did understand his mother needed help. Little Carl grew into a logical man who looked for reasonable explanations. He studied medicine in search of answers. He wanted to understand what afflictions his mother suffered from, but there was one thing he never forgot. His mother was not the only one who saw a glowing figure emerge from her room that night. Perhaps she was unwell but perhaps she was special.

October 1913

Dr. Carl G. Jung, renowned father of analytic psychology, now in his upper 30s, sank into his chair aboard a train bound for Zurich. The seat was broken-in but still had some comfort left in its cushions. He took a deep breath trying to forget the fight he just had with his mentor, Sigmund Freud. [FACT 5]

He looked out the window admiring the passing countryside and thought of nothing more than what he saw in the moment.

The train chugged through baren fields that were either fallow or had recently been harvested. The scene out the window turned from farmlands into meadows and finally scattered trees became a dense forest which dimmed the cabin inside the train; that's where it first happened. Dr. Jung was staring mindfully towards the top of a hill between the trees when a vision abruptly replaced reality.

A great flood swept through the woods. The fields and meadows where cows grazed washed away as water seemed to rise from nowhere. What had been the heart of mainland Europe just moments before, became a raging ocean. In every direction, there was water.

Far to the North he saw a fishing village on the coast. The strange thing was the flood didn't come from the ocean, instead it came from the middle of the continent and swallowed the seaside village whole. Where is this? He wondered. In response he was answered, not by a voice, but by a clear understanding; it is the North Sea.

Where this understanding came from would be one of Jung's favorite topics to contemplate later, but right now his mind was overcome by awful feelings and nightmarish images.

The flood became a maelstrom of destruction churning waves of an awful yellowish gray. In one direction people fled their homes, in another direction they were engulfed by the waves before they knew what was happening. In every direction people suffered. The waves spread far and wide, violently flooding most of Europe. The murky flood waters pushed the whole way to the Rhine River, to the doorstep of Jung's homeland.

The Alps themselves rose to the heavens to stand against this awful watery scourge. The waves battered the mountains like an army charging at an impenetrable fortress wall that protected Switzerland.

The sea lurched again, rubble and bodies were tossed about by the powerful sways. As waves decimated the land, red spots began forming in the polluted water until the whole sea was red. Finally, the bloody flood water began to drain away but instead of rejoicing, the people groaned in newfound agony. The receding waters revealed the full extent of the death and destruction it wrought.

The water continued to lower, and Jung recognized the curves of the Rhine River, which continued flowing red. The water in the river wasn't just mixed with blood; the whole river was blood. What is this madness? Jung posed the question within his nightmare.

A voice in his head rang out, clear as day, "See what is shown and do not doubt its legitimacy, for it is as real as you are."

Impending Doom

Over an hour after the vision spontaneously began, Jung snapped out of it. He was on the train looking at the seatback in front of him. Out the window they were passing through a small Swiss village.

Where am I? He tried to remember but the most peculiar thing happened. For a split second he didn't know who he was, as if his identity had somehow been lost in the flood.

He realized he had been clenching his fists and thrashing his legs in anguish. He remembered he was on a train to Zurich and his personal identity quickly returned. As he looked around other passengers quickly turned away from his gaze. He took a deep breath and looked back out the window to find a perfectly peaceful Europe.

He pondered long into the night on what this vision may have symbolized. A waking dream that caused me to forget my own identity, he thought, it must be a personal message trying to get my attention. There must be some relationship I am neglecting.

Dr. Jung was not a perfect man and suffered the same human imperfections we all do. What changes do I need to make in my life? He thought in frustration.

Jung awoke the next morning with a clear mind. For a few weeks he put the whole thing out of his thoughts and went about life as normal. Until one morning while he sipped coffee at his usual café. He accidently set the mug down too hard creating a loud clang like a gong being struck. The sound reverberated through the café and was just the right frequency to send him into a trance.

The evil flood took over his mind with the same tidal wave of emotions that accompanied it the first time. It swept Europe with just as much suffering and agony as it had when he saw it from the train. Jung was now convinced this psychosis was being caused by something he was doing or neglecting to do. Later that morning he took extra care as he interacted with his secretary. When he returned home for the day he spoke more compassionately to his wife. Nothing he did stopped the recurring vision.

Jung meditated on the meaning of this waking nightmare. He thought about it while eating breakfast, walking through town, and as he laid down to bed. He thought on it so much that as he fell asleep, he dreamt a dream that was nearly identical. That only made him think about it more. He wasn't sure if thinking about it was causing his dreams or his dreams were causing him to think about it.

After months of torment there was one last dream where the voice spoke as plainly as it had the first time, and it finally clicked.

"See what is shown and do not doubt its legitimacy, for it is as real as you are."

This vision is meant to be more literal than I thought. But then what does it mean? He wondered helplessly. Is most of Europe doomed to some awful fate? If so, what am I supposed to do with this knowledge? If this was a message about something I was doing wrong, it would be easy. I could fix what I was doing, and the visions would cease. No, he thought, my heart tells me this is a warning of impending doom. [FACT 6]

FACTS

3.   Dr. Carl Jung, the famed psychologist, recalled an experience very similar to what is described here. One night as a young child, Dr. Jung saw a headless figure emerge from his mother's room. He watched its glowing head float up and away from its body six or seven times only to disappear altogether. He never found a reasonable explanation. In his memoirs, Jung recalled other strange things which, in one way or another, always ended up being associated with his mother or one of her 'odd' relatives. It is a fact that he experienced this event, however, what the explanation of the event is, of course, is up for debate. [Ref 3]

4.   Little Carl's mother was sent to a psychological hospital in Basel, Switzerland sometime after this incident. She lived there for a few months, heavily medicated and highly misunderstood. Jung would later describe his parents' relationship as conventionally right and psychologically all wrong. [Ref 3]

5.   Sigmund Freud considered Jung his successor even going so far as to refer to him as an adopted son. In 1909 they had a disagreement about some of Freud's theories. Jung felt there was more to the human psyche than

sexual metaphor. Freud did not take criticism well and continued to over-sexualize his theories which led to a rift. In 1913 they had a falling out and the strained relationship never recovered. The two men were said to hate each other for the rest of their lives. [Ref 4]

6.   Leading up to WWI, Jung had a series of premonitions. At first, he thought it was some personal message from his subconscious mind, but he came to realize it was a premonition of future events. These visions presented themselves as normal dreams and waking visions. Three months after his final vision, war was declared all over Europe, fulfilling Jung's prophecy.

He later stated that during the time the visions occurred, the idea of a major war in Europe seemed like one of the most unlikely possibilities for the near future. Looking back on the series of events left Jung questioning where the message came from. Thinking about possible answers helped Dr. Jung develop some of his most controversial theories about knowledge existing outside of the knower's physical body. [Ref 3]

Chapter 2

Manufacturing Destiny

This chapter speaks of an energy that sounds like mystic nonsense but as you learn the science behind Quantum Field Theory, described later in this book, you will realize that all physical reality originates as pure energy being hosted or focused by the quantum fields.

"You must name your heir, your Majesty." A Lord counseled the great and powerful Emperor of the Austro-Hungarian Empire. The emperor scoffed as he reviewed his options. His only son met an untimely death leaving his nephew, Franz, as the heir apparent.

The emperor begrudgingly approved. As a child Franz had numerous health issues which the emperor took for weakness. However, it was Franz's sheer determination to overcome his health issues that ultimately shaped his determined and kingly personality.

Franz Ferdinand was given the royal title of Archduke and for the next 25 years he prepared to ascend the throne. He grew to be unlike the monarchs of his time. He planned to include ethnic groups that were marginalized under the current ruling class. Franz wanted all subjects to have their own representation in the federal government, even the Serbs in the South; but he was not the emperor yet so change would have to wait.

His uncle, the current emperor, poured all his might into preserving the monarchy in all its glorious, divine power. He could allow for some progressive changes in the future, but Franz had done something truly unforgivable; he married a commoner. He should have been betrothed for political alliance or land or power, but he found his true love and married her before anyone could stop him. The lovers were so committed to each other they often said they would die in each other's arms.

They had three children whom they loved and cared for. Franz maintained his status, but the royal court treated his family worse than servants. The emperor was furious with his nephew's marriage and felt it tarnished the monarchy. He acted as if God's divinity, which preserved his royal authority, had been diminished by the marriage. The distaste grew like storm clouds on the horizon and for many years the emperor wished he could rid his monarchy of the impurity. That thought, whether intention was with it or not, festered and the flowing energy of quantum creation was swayed by it.

Franz did not concern himself with such petty matters. He was a smart and intuitive man and by May of 1914 his intentions were constantly focusing the energies-of-reality to embody a truly influential leader who would be remembered by the whole world. The sub-quantum energy received those goals and responded with an idea that popped into his mind via intuition. He decided a trip to the Southern region of their great empire would help show the Serbian population there he wanted to include them in the empire's power and prosperity.

The city of Sarajevo, near the border of Serbia itself, was the perfect location not only for his royal duties but also as an anniversary getaway for Franz and his wife to temporarily escape the odious royal court.

June 1914,

Sub-quantum energy swirls in pre-physical parts of reality. Actions in reality pull and push this energy the way a flowing magnetic liquid might sway like a shifting wind around a magnetic pole. At first glance this flowing energy seems mystical but in truth, it is a natural mechanism of physical reality like gravity or electromagnetism.

This subquantum energy-of-creation was flowing somewhere in the depths of reality. At first it was just a trickle but now the energy was like a river being fed by many streams. This energy-of-creation, focused by the quantum fields, created a dingy cellar where candlelight flickered on old stone walls and the faces of men who met in secret there. Though unknown to these conspirators, the energy was flowing from countless millions of people whose future states-of-being were influencing them and the ideas they were conjuring up.

Two of the men were older and respected as mentors. Seven more men were young activists, recruited from local colleges. The oldest man, and leader of the group, had already decided on what they would do. He held the thought in his mind, however, he didn't want to suggest it, for it would not be him doing the deed.

"We must send the empire a message," he said.

"What would you have me do?" One of the eager young men asked almost from a trance. Gavrilo Princip, only 19 years old sat in the back and leaned in to hear the answer with bated breath.

"We must do whatever is necessary to show the emperor he has no power over Serbian people. We, the Black Hand, will carve out a place for Serbs wherever they dwell." The leader encouraged.

The group fell silent. Candlelight danced as the men pondered. The energy flowed into the depths of one young man's consciousness, and he blurted out, "I should kill the Archduke when he comes to Sarajevo next week!"

The old leader stood up and paced the room. "That would be bold," he said. "That would show, not only the emperor but all Serbians, we are not to be underestimated. We have a right to govern ourselves!" The man shouted passionately.

"You would kill the Archduke," the other older man said in a whisper, quieting the room. "How?" The simple question posed to the group set off a shockwave in the realm of that ethereal energy which connected their hushed conversation to countless people all over the world. The quantum energy flowed like alternating current; forward and backwards, simultaneously influencing and being influenced by the men in that room and the millions upon millions of unknowing participants all over the world who would soon be directly affected by what they were planning.

The young men, unconsciously swayed by the desire to gain the respect of their mentors, allowed themselves to be overcome with the idea of glory for their people against a tyrant. They thought desperately about how they could do the deed. The energy flowed through their consciousness producing ideas that seemingly came from some recess of their brains but really came from somewhere outside of them altogether.

One young man took a deep breath hoping the answer would reveal itself and release his growing anxiety. As if someone else was speaking, he heard himself say, "I will shoot him."

Gavrilo heard those words and was overcome by a feeling that started in his heart and sank into the pit of his stomach. It was an emotionless understanding that even though it was spoken by someone else, it was his truth, his future, his destiny.

"I can get a bomb." Another said as the thought popped into his mind.

A third remembered an acquaintance who mentioned he would be setting up for the Archduke's motorcade. "I can find out what route his car will take." He said as the plan took shape.

The old man began speaking louder, "Because of you the whole world will know the Black Hand. The emperor will learn the might of the Serbs and we may yet have Balkan independence." He raised a defiant fist to the air, "Serbs everywhere will realize their power because of you!"

Gavrilo looked at his co-conspirators as they got carried away. He looked into their faces and saw excitement but also fear. The primordial energy surrounded Gavrilo before flowing through his heart, pouring into every cell in his body. It will be me who kills him, he knew. He didn't cheer or proclaim he would accomplish great deeds. Instead, he remained silent as the feeling of a great weight was laid upon him. Beyond any doubt, he knew it would come to fruition.

After some discussion the plan was set. Seven young members of the Black Hand reviewed the planned route of the Archduke's motorcade. The Serbian government provided explosives and pistols, and Gavrilo Princip was assigned as the backup assassin to kill the heir to the Austro-Hungarian Empire.

On June 28th, 1914, seven assassins lined both sides of a street named Appel Quay which overlooked the river gently flowing through Sarajevo. They blended in with locals who stood as if a parade were about to begin.

A motorcade of six cars turned a corner and unknowingly drove toward the attackers. The Archduke sat conspicuously in the back of a convertible next to his wife. As the car approached the kill-zone the primary assassin momentarily locked eyes with the Archduke. He pulled the pin on the grenade, and as the car passed, he threw it, aiming for the back of the Archduke's head. The motorcade chugged on, and the grenade fell short, hitting the back of the car before exploding underneath the following car as they drove over it.

The explosion injured the military officers in the vehicle but did little more than startle the Archduke and his wife. Gavrilo Princip watched from across the street as the Archduke chugged away and bystanders grabbed the conspirator who threw the grenade.

"Why would you do that?" An old man shouted at the grenadier. "He wants to help Serbian people!" An angry mob surrounded the man; they wanted revenge. Their combined intention churned the energies of quantum creation to capture the assassin, but the man who threw the grenade had other plans.

Princip thought for a moment that he should run across the street and assist his friend's escape, but something held his feet firmly where he stood. His friend shook off the angry mob and reached into his coat pocket pulling out the cyanide capsule his handler distributed to each of the assassins. He ran to the edge of the street and looked to the river flowing in the canal below. He leapt onto the railing and shouted, "For Balkan Independence!" He swallowed the cyanide and jumped into the river.

The people who had been grabbing at his coat watched as their attempts to catch him failed and he seemingly traded capture for death.

The energies that create physical reality flow in directions we cannot comprehend. The intentions of the angry crowd rang out as ripples of very real energy into the 5th, 6th and 7th spatial dimensions. [FACT 7] This reality-generating-energy manifesting the grenadier to his death abruptly changed directions like a river that began flowing backwards.

The would-be assassin fell to the water below but instead of a splash into a quick death, he hit the bottom of the riverbed, at that moment realizing the water was only four inches deep. No bother, he thought as he lay in the cool stream waiting for the cyanide to take him. The angry mob gathering above became determined exact revenge. The energy responded.

Death never came for the would-be assassin. Instead, he began to wretch. An awful ache pounded in his head as pain from the cyanide wreaked havoc through his body. Minutes later police apprehended him. The cyanide pills were too old to do anything other than cause agonizing pain.

Gavrilo had known all along the first attempt would fail. He didn't know how but the feeling in his heart had already warned him, this was his task. If the grenade attack failed, his orders were to turn away from the river and walk one block North to wait along the Archduke's planned exit route. Instead, he found himself walking along the river in the direction the motorcade had traveled. He walked up a block, crossing a road before his feet turned left and another half dozen paces up a side street, away from the river.

He stopped and looked back at the street signs. They read, Appel Quay and Franz Joseph Strasse. His orders were to wait along Franz Joseph Strasse a block and a half West, but his feet would not budge from the spot where they delivered him. He stood near that intersection as if in a trance. He felt no hunger nor summer heat. He remained calm as the energy of quantum reality flowed through him.

Meanwhile the Archduke arrived at the Town Hall as planned where he proceeded to berate the local Governor as if the bombing was his fault.

Manufacturing Destiny

Before leaving the Town Hall, he heard the motorcade drivers talking in German, which was the common language throughout the empire. They discussed the planned route out of the city, along Franz Joseph Strasse. The Archduke corrected his drivers; he was going to the hospital to visit the victims of the grenade attack. His security objected to no avail and his drivers discussed a new route. They agreed to drive straight there via Appel Quay instead of turning down Franz Josef Strasse as if they knew an assassin was supposed to be there. The Archduke approved and it would have worked perfectly, had it all gone as planned.

The primordial energy that perpetuates reality directs all things, even retroactively. Earlier on that fateful day the driver of the first car became ill and was replaced by a driver who only spoke Czech, not German.

The convoy departed with the intention of arriving at the hospital but that potential state-of-reality would never come to fruition. The Czech-speaking driver in the lead vehicle followed the original route since the change was only discussed in German. He turned off Appel Quay onto Franz Joseph Strasse, followed immediately by the car chauffeuring the Archduke and his wife.

The driver of the second car quickly recognized the mistake and abruptly stopped. He honked his horn to indicate to the first driver he had erred. The blast of the horn rang out into the street and reverberated in the eardrums of none other than Gavrilo Princip standing 10 feet away.

The driver quickly put the car in reverse as Princip stepped toward the stalled vehicle, drawing his pistol. The driver put his foot on the accelerator just as Princip closed his eyes and turned his head. He prayed, guide my shot and let Serbians be free from this empire.

The car began moving in reverse as the assassin fired two shots and the energy-of-ultimate-creation caused the bullet and the Archduke's neck to occupy the same physical space at the same moment in time. Objective reality experienced the assassination that would create the modern world.

The first shot hit his wife, Sophie, in the chest, the second hit the Archduke in the throat. The lovers died in each other's arms on a street named after the current Emperor as if he had orchestrated the whole series of events. Princip turned the gun on himself but was thwarted by a passerby who tackled and beat him until the police took him to prison.

"Your Majesty," said a servant to the emperor, "Your nephew and his wife were assassinated in Sarajevo, I am so sorry."

The emperor stopped in his tracks. He was stunned but an odd feeling of relief was radiating from his chest. The emperor thought of his nephew and felt sorrow for the man but he knew all too well the power of a monarchy can be strengthened in chaos. The emperor had heard rumors of assassins in Sarajevo but allowed his nephew to proceed anyways.

The emperor felt like he sacrificed his nephew. He would use this as an excuse to gain power over Serbia. He knew he could weave the tale any way he wanted to mass even more support for his rule. Possibly best of all, he knew his monarchy would remain pure-blooded as his nephew's commoner wife had been killed by his side.

He knew he was a powerful man, but he had no idea that his intentions and his knowledge of how to affect change had contributed to the same pool of energy that generates reality itself. He had no idea that he had assisted in directing that energy to this single event and to a great war that would change the world forever.

The people of the Austro-Hungarian Empire were outraged, declaring War on Serbia a month after the assassination. Russia began mobilizing its military to honor its alliance with Serbia. Germany came to the aid of their close friends the Austro-Hungarians and because of Russia's alliance with France, Germany automatically declared war on France as well.

The energy that reciprocates our intentions into reality is absolute. Sophie and Franz were the first casualties in the Great War that would create millions of tragic love stories. The final casualties, however, would be the monarchies themselves. Kings and queens, emperors and czars began the war but ended up powerless because of it.

Every individual has just as much primordial influence as any other. Kings and slaves all contribute their consciousness to it. By 1914 it was high time for the ultimate-power-of-creation to transfer from select individuals claiming divinity to all people. There were other ways the primordial energy could have achieved this end-state, but war was the path we unknowingly chose.

For this great coincidence to occur, all participants' deepest intentions held true. The Archduke was a great leader and became immortalized, though only for his death. He and his wife died in each other's arms as passionate lovers. The emperor's monarchy was rid of impure commoners and he used the event to gain power and influence. The Serbian assassins sent the harsh message they desired. The people got the satisfaction of catching the assassins and beating them before they went to prison. As they originally intended, the assassins became martyrs after later dying in prison. The influence of the last participants is the most intriguing. It is the indirect participants only affected by second and third order affects. Be specific what you wish for, you just may get it.

This singular event directly caused World War 1. World War 1 caused monarchies around the world to be stripped of their divine power, just as the Black Hand intended. This created a more democratic world for, now, billions of people. World War I led to World War II, the rise of globalism, many technological advances in medicine and energy, and much of the modern world. The World Wars created countless tragedies in millions upon millions of individuals' lives. In one way or another, almost every living person since that event has been affected by it. [FACTS 8, 9]

FACTS

7.   Leading theories in modern physics require (String Theory) or at least hint at (Quantum Field Theory) the existence of more than three spatial dimensions. These spatial dimensions are only accessed by the very smallest bits of physical reality. There is a more detailed explanation in later chapters.

8.   As strange as it may sound, pre-physical energy is creating every single event, action, thing, and perhaps even thought, as we individually and collectively experience reality.

The primordial energy described in this chapter will be explained in scientific detail in the latter half of *The Owl and The Osprey*.

9.   This was an accurate depiction of the circumstances surrounding Archduke Franz Ferdinand's life and assassination. The car with the Archduke did stop, of all places, immediately in front of the assassin after taking a wrong turn. There is no evidence to support the popular claim that Gavrilo Princip wandered off looking for a sandwich. There is no evidence that Princip was not anywhere other than the exact location he was directed to be in, or that a driver of the motorcade was replaced at the last minute, however it is believed the wrong turn was caused by a language barrier of the lead car's driver. [Ref 5]

        The ultimate law of creation hides in plain sight. The Archduke was assassinated in a car bearing the license plate "A III 118" as if this seminal event was simultaneously both-ends of the Great War. WWI ended over four years later, on Armistice Day 11/11/18, or A111118, just as the license plate suggested. This went undetected until 2004 when historian Brian Presland noticed the odd coincidence. [Ref 94]

Chapter 3

Dispatched into Darkness

January 1915,

    The cold night air slashed Rupert's face as if he were flying through a field of tiny razors. He sat in the cockpit of his biplane, sandwiched between the upper and lower wings, gliding through the night sky. His plane was a simple assembly of framed wood wrapped by a thick canvas. Cables ran from the controls to the various wing sections just like brakes on a bicycle. Rupert pulled back on the mechanical controls and a cable went taught, folding the back edge of the wings upwards. He gained altitude but could barely see anything in the darkness besides a few lights from the air base below. In the sky, he was all alone. To the West, the dim glow of the setting crescent moon faded, leaving only the stars shining in the night. They should be here by now, he thought impatiently.

    His brown leather jacket and tight helmet helped protect him from the cold, but it was the adrenaline in his veins that kept him warm. All the aviators were issued white silk scarfs; Rupert's was tucked snuggly into his wool-lined jacket and his goggles were pulled over his eyes. Without goggles, it would be impossible to see anything since his cockpit was open, leaving him exposed to the air and elements.

He checked his compass and verified he was on the correct heading. He looked back into the darkness and was instantly blinded by a bright light right next to him. He swerved down to the left and saw the underside of another plane swerve up in front of him.

"What am I doing?!" Rupert yelled out into the night. He quickly turned on his flashlight and frantically waved it around until he spotted the other planes. He signaled to them, and they signaled back before settling into formation around Rupert.

Just 20 minutes ago, the air base received reports of some flying monstrosities terrorizing a village on the coast. No one recognized what the giant flying beasts were, but they rolled in like a mechanical thunderstorm. The horrendous noise of their wailing engines spread fear into every nook and cranny of the town they were bombing.

Rupert and the other pilots were quickly dispatched to intercept the beasts. The four biplanes raced through the darkness for over half an hour before light from the coming dawn started reflecting around the atmosphere directly in front of them. Stars faded from sight as a deep blue gradually pushed the black night into the sky behind them.

The war had been in some far-off land for such a long time. The people of England felt removed from it, like it would never directly affect them. All that changed when the German's unleashed their new airships in this unprecedented attack. Now, in the darkness, as explosives rained down and the people helplessly shrieked into the night, they realized they had been part of the war all along. The Great War finally reached English soil. [FACT 10]

First Lieutenant Rupert Robinson and his men were members of the Royal Flying Corp assigned to the 39th Home Defense Squadron stationed at a makeshift air base called Sutton's Farm. They were ordered to find the flying machines, gather intelligence, and engage the targets if possible. [FACT 11]

Rupert was the Flight commander, leading his men to meet this unidentified foe head on. The sun finally peaked over the horizon and the Eastern sky looked like a grenade of light exploding across the heavens before fading to a pale blue. The village came into view and the sparse, hazy clouds overhead looked like the final strokes on a painting of the idyllic English countryside.

Rupert was not concerned with the sunrise; it was the little black specks in the distant sky that had his full attention. His wingmen saw them too and instinctively tightened the formation around their leader, racing to defend their homeland.

Hand grenades and bombs continued to rain down on the town. At one point a German fired a machine gun sporadically into the buildings. People cowered in corners hoping it would end.

The wind and the rumble of the biplane's engine normally forced the young pilot to yell very loudly to his observer, William, seated immediately behind him. This mission, however, was different. The squadron commander made a last-minute decision to leave all the observers on the ground at Sutton's Farm. With less weight the planes would burn less fuel and would be able to fly farther.

Finally, the planes soared over the seaside village that had endured the raid. In the morning light the attackers were making their escape over the North Sea. Rupert increased his speed, pushing his plane to the limit. Beneath them, the pleasant green countryside ended abruptly, giving way to a white-capped ocean, glistening in the morning sun.

Rupert stayed focused on his mission, only breaking eye contact when he noticed the waves below bashing into the rocky coast, like relentless Viking raiders. His gaze quickly returned to the silhouettes in the sky ahead. Two giant ovals became visible; one dull grey and the other light brown, both seemed to effortlessly float across the sky like unholy clouds in a storm sent from hell itself. The sight of these abominations gave Rupert a queasy feeling in his stomach, but he led his flight forward.

"Zeppelins." Rupert mumbled to himself in his empty cockpit.

The four biplanes closed in, splitting two planes for each giant blimp. They buzzed circles around the massive aircrafts but all they could see was canvas. Rupert flew underneath the grey one and spotted the cabin. Without his observer he was forced to gawk into the enemy ship's control room while holding his plane steady. He got so close that for a split second he made eye contact with one of the Germans who appeared to be waving playfully at Rupert.

Rupert had never been so angry in all his life. He pulled up and circled back around for another pass. These Germans just bombed civilians, he thought, now they were playing games as if this was all a big joke.

The biplanes chased the zeppelins far over the ocean. Rupert leveled off, then turned in a downward curve coming so close that the tip of his wing grazed the canvas of the blimp. He steadied his plane with one hand and drew his pistol with the other; it was the only weapon he had.

His anger lingered, but then he felt something new. It was the most peculiar sense of clarity; it was almost comforting. All the worries in his life were back on the ground at Sutton's Farm. He was laser focused on his mission. He didn't even have to think, he just let his training kick in and his body did everything it was supposed to do.

This time as he circled passed the cabin, he fired three shots from his revolver. For the first time in his life, Rupert was free from his own morality. He felt alive. It was kill or be killed, there were no rules, just survival.

Shaking with anger and excitement, he holstered his pistol and grabbed the controls with both hands. He wondered whether he hit anything and was answered by a faint cracking sound; over and over, each pop was like the crack of a whip taming a lion.

I hit it, he thought in excitement, that must be the blimp crumpling in on itself. He instinctively flew away from it, figuring it would explode any second. His head was filled with the idea that he must have somehow fired the perfect shot. Surely, he would be decorated with awards for bringing down a zeppelin with nothing but a pistol. A wonderous feeling radiated from his heart. I will be known far and wide as the protector of England, he thought. The King himself would knight me; Sir Rupert Robinson had a nice ring to it. He waited for the explosion.

The only thing to explode was the tail rudder of his own plane, and with it, the idea of him being knighted by the King went up in smoke. He turned to see what happened and quickly realized the cracking sound was a machine gun the Germans were firing at him. Bullets whizzed past his head like a swarm of angry wasps. How reckless of me to get lost in a silly daydream, he thought. Apparently, he missed his mark, but the Germans were quick to react and unfortunately for Rupert, they didn't miss theirs.

Dispatched into Darkness

With a damaged tail, it would be nearly impossible to maneuver his plane. Rupert wrestled the controls and finally got turned towards the mainland. He pulled up and felt a sharp pain in his right arm. He realized he was injured, either by the machine gun or by the shrapnel created from the splintered tail of his plane.

More machine gun fire rang out, but it sounded different. Rupert turned to see the only biplane in his formation equipped with a machine gun was firing wildly into the side of a blimp. The zeppelin endured and was now so far out to sea they had to call off their counterattack.

I'll never make it back to Sutton's Farm like this, Rupert thought, trying to keep his plane steady with his good arm. He visualized all the procedures for a normal landing, noting the changes he would have to make because of the damage. He gained a little altitude to give himself some extra time to react in case he lost control. [FACT 12]

Rupert looked from his compass to the horizon, making a small course adjustment.

"Steady," he said to himself, turning slightly left. There was a loud crack, and the top half of the tail broke off the plane and fell to earth.

"Too far!" Rupert yelled to himself. "It must have been hanging on by a thread."

Without a tail rudder Rupert worried that making any adjustments might force his plane to spin wildly out of control. He was now at the mercy of the air beneath his wings. He looked ahead and saw the waves pounding the rocky shore.

The adrenaline started to wear off and he became acutely aware of the pain in his arm. He glanced at it and saw a large rip in his jacket. There was blood but he couldn't differentiate shadow from shrapnel. He looked again to the coastline and wondered if he was going to crash into the rocks just like one of the waves. That's when a terrible thought took over his mind.

Why did I notice these rocks before? All my attention focused on them. They had nothing to do with my mission, my eyes should have been fixed on the zeppelins, but I was drawn to the rocks. Are they my destiny? Rupert thought, allowing the despair to corrupt his mind. I'm losing altitude. He looked back and saw smoke billowing out the back of his plane. This is it. My destiny is to die on those rocks.

"One cannot escape their destiny," he said bleakly. He went numb and let his hand slide off the controls.

In that dark moment, a wingman caught up to Rupert and signaled for him to keep going. The wingman animated his hands trying to communicate something. Rupert's good hand grasped the controls and, without thinking, he gently pulled straight back and raised his plane ever so slightly.

Looking down, Rupert realized he could make it past the rocks. He spotted an empty field and shook the feeling of impending doom. The field has been in plain sight this whole time, he laughed to himself while grabbing his controls as tightly as he could, preparing for a rough landing.

With the morning sun at his back, he had a good view of the field and the flat strip of land he was aiming for. He eased his plane lower, his arm still throbbing in pain. Just hold on, he thought. Still lower he flew, now barely hovering above the ground. His wheels touched down and the plane let out a crunch that sounded like bones breaking. It bounced and pulled hard to the right; a wheel must have given out, he thought.

Rupert pulled the lever for the brakes. He let go of the other controls and pulled with all his might. His remaining wheel hit a bump, jerking the plane off the makeshift runway. It bounced up throwing Rupert off his seat. The plane crashed back down hard, flinging mud into the air, finally skidding sideways to a halt in a farmer's empty field.

The other pilot landed across the field. He wrapped Rupert's arm in his white scarf and helped him into his observer's seat while the other pilots circled overhead. He suffocated the small fire on the tail of Rupert's plane and took note of the coordinates where they were leaving it. They returned to their base at Sutton's Farm and got Rupert the medical care he needed before reporting to their commanding officer.

FACTS
10.   On January 19, 1915, German zeppelins attacked the English mainland for the first time. By the time the war ended they would conduct over 50 bombing raids in Great Britain. [Ref 6]

11.   The character Rupert is based on real life Lieutenant William Leefe Robinson. After his initial training, Leefe was sent directly to France. There, he flew range-finding missions where he would scout out German targets and report their locations to the Allied artillery. While flying a mission there, he was struck by shrapnel in his arm.
       The injury was cause for his reassignment to the 39th Home Defense Squadron located at Sutton's Farm Airfield, back in England. There, he received training for night flying and was able to recover from his wounds. [Ref 7]

12.   Modern pilots use a visualization technique as a rehearsal before performing in air shows. This visualization is done as a group with all the pilots closing their eyes and imagining they are in their planes. This meditative form of visualization, also used by professional athletes, is the first step in the law of attraction. [Ref 8]

Chapter 4

Angel's Glow

"Truth is stranger than fiction, because fiction is obliged
to stick to possibilities; truth isn't."

-Mark Twain

William, Rupert's observer, ran to the edge of the runway as the
planes returned. He, and seemingly everyone else at Sutton's Farm, watched
Rupert get carted off to the infirmary as soon as the plane came to a stop.
William followed straight away but froze at the sight of his good friend's
injury. Rupert's face was relaxed; he must not be in too much pain, William
thought.

"You finally get your chance, and you go off and get all bloodied up."
William joked. "And you went and lost *my* plane?"

"*Your* plane!?" Rupert said, comforted that his friend was there for
him.

"Only joking mate, I'm glad to see you back in one piece." He
glanced at Rupert's bloody arm, "Honestly, I'm jealous, now that you've seen
combat and have a scar to prove it, you'll have an easy go with the ladies
down at the old pub."

"No pubs until this arm is fully healed!" The nurse ordered as she
wacked William with a roll of bandages.

The next morning Rupert left the infirmary, all patched up. At first, he barely noticed the pain, however, days turned into a week and the gash showed no signs of healing.

"Come on, let's go for a nice easy jog." William said one morning.

"I can't." Rupert said.

"I'll go easy on you."

"My arm is killing me; I won't make it 10 feet."

"You need to get that sorted, mate." William encouraged. "Maybe it's not healing because you're not using it. A jog might help."

William left the makeshift barracks alone, for a quick run around camp. The whole base had been constructed in a hurry when the war started. Their barracks were nothing more than plywood walls and loose planks all slapped together with as many beds shoved inside as they could fit.

It had a dirty feeling that stuck with them, no matter how long of a shower they took. It was uncomfortable but it was better than the trenches on the front lines.

"What are you doing?" William asked as he sat down on his bunk after his run.

"I'm changing the bandage, it aches horribly."

"Aw, it stinks too!" William said as he chased away a pair of flies from Rupert's festering wound.

"Back to the infirmary for you lad." William said sternly.

Rupert was in too much pain to resist.

"You should have brought him in sooner!" The nurse scolded William.

"Me? He's the one that said he was fine!" But William's words fell on deaf ears.

Turning to Rupert the nurse put on a pleasant motherly tone, "Let's get you cleaned up, Dear."

"Is there anything that will ease this pain?" Rupert asked through a grimace as the nurse cleaned his wound and applied fresh bandages.

"Keep it clean, get some rest and it will get better." The nurse said bleakly. William noticed a look on her face that concerned him. There was something she felt but wasn't saying.

Angel's Glow

It was another exceptionally cold night, but Rupert's arm became uncomfortably hot. He took off his bandages and allowed his arm to breath while he slept. William was in his own bunk but couldn't sleep. He was thinking about his arch nemesis who humiliated him, gambling over a dice game earlier that day. He laid in his bed imagining what he would have done if it had come to blows; he threw a right hook and then an uppercut. After finally getting even in his imagination, William felt better. He took a deep breath and rolled over to go to sleep but before closing his eyes, he saw it.

A dull, but very real, blue glow was coming from Rupert's bunk. He rubbed his eyes, trying to focus on whatever this thing was but it didn't have a shape. He looked around the quiet room in disbelief, he had never seen anything like it. He crept through the darkness over the squeaking floorboards until he was right next to Rupert. William couldn't believe what he was seeing. It was the gash in Rupert's arm; it was glowing.

"What is this madness?" William whispered in distress, not realizing he was speaking out loud.

"What?" Rupert stirred.

"Mate? Your arm! Are you alright?" William asked with true concern in his voice.

"What are you talking about, I'm asleep." Rupert mumbled as he rolled over.

"Rupert, your arm! It's glowing!"

Rupert lifted his arm into view and saw it. He bolted upright in his bed, now very awake. The pair excitedly hurried to the infirmary where William happily woke the nurse.

"What's the emergency?" She asked impatiently, still half asleep.

"Well, I wanted to bring him in as soon as possible ma'am, not like last time!" William never missed an opportunity for a cheeky comment.

"What's happening to my arm, Nurse?" Rupert said in a distraught voice.

The nurse stared in awe as she inspected the wound. "I've never seen this before. Your wound is glowing! What happened?"

"Nothing! I went to bed like normal and then he noticed it."

"Does it hurt?"

"No. If anything, it feels better."

"It looks better!" The nurse said, surprise, as she inspected it. "I don't know what to say. Go to sleep and come back first thing in the morning. If it starts hurting come back sooner but for now let us thank God for sending you an angel."

Rupert laid in his bunk staring at the blue glow on his arm. It was mesmerizing. What is it? How did it come to be? I feel like I have been chosen, he thought. As Rupert drifted into a gentle sleep, he imagined a heavenly angel dressed in white, floating overhead, protecting him.

The news of Rupert's glowing wound spread around Sutton's Farm. Naturally, everyone wanted to see it. William, ever the showman, instructed Rupert to cover his arm with a cloth and when the time was right, he would dramatically pull it off in a big reveal. Dusk turned the twilight sky into night while William held the crowd outside Rupert's barracks. When William decided it was dark enough, he silenced the crowd and knocked on the barracks door. Rupert emerged to the restless crowd.

"It's covered up." Said an impatient voice.

"Gentlemen, please give me your attention." William announced dramatically. "In just a moment you will all be witness to nothing less than a miracle from God himself. Many of you know, I discovered this miracle because I was taking such good care of my closest friend, trying desperately to nurse him back to health. Last night, I was lying awake praying for Rupert, as I do every night and finally God answered my prayers! One might say that I actually caused this miracle, but I humbly disagree."

"Get on with it!" The same impatient voice interrupted as if he had a stake in the matter.

"Alright, alright." William smirked. "I don't want to keep you waiting any longer." He grasped the cloth draped over Rupert's arm and counted, "Three, two, one..." He pulled off the covering and the wound glowed in the night. The crowd burst into excitement.

"Would you look at that!"

"Do my eyes deceive me?"

"Praise be to God!" Someone else exclaimed.

"Out of the way, I can't see it!"

Amongst the noise of the crowd Rupert heard William talking to his arch nemesis.

"Get a good look Smith! Go on, take your time," he said smugly. "Turns out I wasn't lying after all, huh? I'll be right here when you're ready to pay up."

Within days the wound went from a festering infection to almost completely healed. This mysterious glow was clearly the reason, but no one understood what it was. The nurse asked doctors and every medical professional she could, but no one knew. Some doctors even began questioning her sanity, so she stopped talking about it altogether.

No one had ever heard of it before, and no one has ever experienced it since. Sure enough, the very real tale of the Angel's Glow faded from fact to folklore. [FACT 13]

FACTS

13.   Angel's Glow was a real phenomenon. During the American Civil War, at the battle of Shiloh, some wounded soldiers did, in fact, have wounds that glowed light blue. The soldiers with glowing wounds were more likely to survive and healed quicker so they called it Angel's Glow. With no cameras, telling the story was the only way to record the phenomenon so, naturally, it became folklore for 139 years.

It turns out the glowing was caused by a very specific bacteria called Photorhabdus Luminescens. It requires extremely precise conditions which have only ever been met naturally at the Battle of Shiloh in the 1860's. The temperature, humidity and bacteria's food source all lined up perfectly for this bacterium to be produced naturally in a human wound. There are no other known instances of this occurring naturally throughout all of recorded history.

What's more interesting is the scientist who proved it. It was seventh grader, Bill Martin who visited the Shiloh battlefield and read about the Angel's Glow folktale. Realizing it might have a scientific explanation, he and classmate Jonathan Curtis chose this topic for their submission in a high school science fair. [Ref 9]

Which old wives' tale will be proven true next?

Chapter 5

Expecta Dominum

September 1916,

"Another patrol? We just did one last night!" William complained as he climbed into the cockpit behind Rupert.

"We must stay vigilant, William." Rupert said proudly.

"I thought we'd be enjoying a pint this evening or playing some Crown and Anchor."

"No need to waste your time on a silly dice game." Rupert instructed as he flipped a switch on his dashboard.

"Silly?" William exclaimed, "I won three pounds off that big bloke last week!"

Rupert signaled to a crew member in front of the plane who spun the front propeller causing the engine to burst into life.

"Not to mention the extra attention I get when I'm winning. There's a fit lass I've had my eye on and one of these days I'll be introducing her to my birthmark, if ya know what I mean." He pointed to a spot on his hip where he had a very unique birthmark, shaped like a plus sign with an arrow above it.

"And you call it silly." William continued, "You're the silly one, always wanting to go out on patrol! You know you're not going to save all of

England by yourself." William's voice was drowned out by the rumble of the engine as Rupert drove the plane forward.

They took off at twilight for a routine four-hour patrol. If Rupert ever got the chance to go out on a night mission, he would. He always came back extra tired and slept like a rock. On nights he didn't get to fly, Rupert was often tormented by nightmares of zeppelins or crashing his plane, forcing him to relive his encounter. In some dreams he didn't make it back to the field and woke up in a cold sweat just as his plane crashed into the rocks on the coast. On more than one occasion Rupert screamed out, waking the other aviators who all shared the large open room.

"What's wrong mate?" William asked the next morning at teatime in the briefing room. "You've hardly touched your tea, and these are the best stale biscuits at Sutton's Farm." He picked one up and tapped it on the table. The biscuit sounded like a rock, "Don't want them to go to waste, now do ya?"

"Not much of an appetite, I'm afraid. Maybe it's this weather." Rupert replied.

William leaned in close, "I don't think it's the weather, mate. You know you really should talk to the Chaplain."

"William, I do appreciate your concern but there is nothing to discuss."

"Well, that's just it, you don't have to discuss anything. Just spend time with him. I know! You can talk about the Bible; you love the Bible." Said William grasping at the fact that Rupert occasionally read from a small bible he kept in his trunk.

"I don't know what good that would do." Rupert said, brushing the whole thing off as he started another round of Crown and Anchor.

When weather was bad the aviators stayed in their respective briefing rooms until the storm cleared. The briefing rooms and mess hall were made of the same cheap plywood their barracks were. None of the buildings were finished properly; seams did not always line up, leaving gaps in corners and edges. Living like that made the men stationed there feel incomplete, but for the time being, it was home.

The briefing room was a big space with a few small offices lining one side of the building. There were diagrams of planes and large maps of the area that included everything from London to the English Channel.

"With the way this weather is, I doubt we'll fly today at all," one of the pilots said calmly, looking out a window across the room.

BAM! The door flew open and hit the wall. Someone was coming in, but a gust of wind caught the door, flinging it open. All the aviators flinched and sat up, one even jumped to attention thinking it was their commanding officer. Instead, in walked a tall skinny man with mud on his boots and around the ankles of his trousers.

"Well, that's one way to make an entrance." The man announced as he pushed the door closed against the wind.

"Oy! Mornin' Chaplain!" William said from the table closest to the door.

"Yes, a mighty fine English day it is. You boys are staying out of trouble, I see!" The Chaplain said as he brushed rain off his large overcoat.

"Care for a game of dice, Chap?" Rupert offered.

"Actually, Leftenant, I was hoping to talk to you privately."

Rupert was shocked until he remembered the conversation with William. Rupert shot William a questioning glance, but he shrugged and looked just as surprised as Rupert was. Rupert nodded to the Chaplain and excused himself from the group to find an empty office to speak in.

"Rupert, I was speaking to your commander yesterday about some other matters and your name came up. He told me you have volunteered to go to France to replace some of the pilots who…can no longer fly."

"Yes, that's right," Rupert replied, a bit relieved he wasn't asking about the nightmares. "Almost every officer has volunteered."

"Right, but um," the Chaplain took his time to find the right words. "It seems, your case may be different than the other officers. Do you ever get woken up at night, Rupert?" He asked, abruptly changing the conversation.

Rupert couldn't help but feel betrayed. But then again, he thought, William was surprised too, and the chaplain has been out visiting other units for weeks. Perhaps it hadn't been William, but if it wasn't him then how did the Chaplain find out about his nightmares? Did his commander know as well?

"Yes, I suppose I've had a few nightmares," Rupert began, "but they are just bad dreams, nothing that would affect my performance in the air. And what of it anyway, everyone has nightmares from time to time."

"Rupert, I am not concerned with your performance as a pilot. I am the Chaplain; I am concerned with your soul."

Those words pierced Rupert's defenses, but he did not know what to say so he stayed quiet.

The Chaplain continued, "All the pilots were asked if they would volunteer to go to France and everyone said they would. You, on the other hand, are asking to go. You asked your commander 18 times in the past month. He told me he's been counting ever since you asked him twice in the same day." [FACT 14]

"So, the commander sent you to talk to me, did he?"

"Actually no. I was talking to some other pilots, and they mentioned someone has been screaming at night waking everyone up. My question to you a moment ago was not about nightmares. I asked if something has been waking you up. I expected you to say someone was screaming at night. You inadvertently told your own secret. I have to wonder, Leftenant, is the reason you so badly want to go to the front somehow related to your nightmares?"

Rupert's defenses went right back up. He felt like the Chaplain tricked him. "It's true, I do have nightmares, but I don't see the point in talking about them, they are just dreams."

The Chaplain heard the frustration in Rupert's voice and allowed a deafening silence to ring through the office before gently saying. "Can you tell me about the dreams?"

Rupert first told the Chaplain about his real experience when he was shot down by the Germans and crash landed in the field. He explained that the dreams were the same event over and over again.

"I don't know why I want to go to the front so badly." Rupert was finally honest with the Chaplain. "Perhaps it's that if this war is coming for me, I just want to get it over with. The waiting in angst sometimes seems worse than just doing the thing you're worrying about."

The Chaplain understood that anticipation was difficult but he could not empathize with the desire to go to a war where so many young men were dying. "Yes, it can be difficult trying to figure out what God has planned for us."

Our destiny, Rupert thought.

The Chaplain felt like he needed to offer Rupert some guidance, so he said, "Do not be so quick to go to war."

Rupert was absolute in his position. "Chaplain, this is a classic tale of good versus evil. They are attacking our homeland and I must rise to defend it."

"Does doing evil deeds for a good cause make them less evil?" The Chaplain challenged.

"It makes the actions righteous." Rupert said quickly.

"You are righteous with your logic, for now. But what if you end up attacking their homeland? Then will they be justified in defending it by fighting you?"

"No! They started the war in the first place." Rupert retorted.

"A young man on the other side might not see it that way. Not once you are flying over his home and he never saw what they did to yours. All I am saying is, be careful what you wish for, you just may get it." The Chaplain understood that fighting in a war changes a man, he genuinely wanted to protect Rupert from becoming someone who committed evil deeds, even if the cause was justified.

There was an awkward pause until the Chaplain spoke again.

"I know that might not ring true to you now, but here is something from Psalms that I think will. It says, wait for the Lord, act manfully and let your heart be strengthened and support the Lord." The Chaplain recited.

"I know this verse!" Rupert blurted out, "Expecta Dominum, it's the motto of the school I went to back in St. Bees!"

"Well, no wonder I thought of you when I read it. I'd say you are definitely acting like a man; you are honorable, and you want to do your duty for King and country. But are you willing to wait for the Lord? We cannot always know why He chooses certain things for us or the timing of those things. Sometimes all we can do is pray and have faith in His plan for us."

"Our destiny." Rupert mumbled. "Chaplain, can we change the plan God has for us?" He asked abruptly.

"All you can do is pray and have faith that He will deliver you from evil." The Chaplain's personal sermon comforted Rupert. It reminded him of when he was a young teenager sitting in the pews of the small church in his seaside village in Northern England. The sun began shining through the window as the conversation ended, just in time to get a quick patrol in.

Rupert's mind raced with thoughts looking for clarity. For the first time ever, he stopped to think about why he was so desperate to go to the front. Flying alongside his wingmen in the pale light of an overcast dusk, he thought back to the fateful night when he crash-landed in the field.

Rupert realized that waiting for combat was more nerve racking than actually being in combat. He remembered that peculiar clarity he felt, like nothing else in the world mattered other than what was happening around him in that moment.

Is that why I want to go? He wondered. Am I just trying to get something over with when I might not have to go at all?

The next morning Rupert, William and two other aviators were called to their commander's office.

"Congratulations, men." The squadron commander said sternly as they stood rigidly at the position of attention in his office. "I just received orders for you four to report to France. It looks like you'll be seeing some action after all. You leave here in one week."

Rupert was flooded with emotion. Out of all the thoughts and feelings he had, one stood out; he wasn't sure he was getting what he truly wanted, but he was getting exactly what he asked for.

FACTS

14.　　　Leefe Robinson pestered his leadership about going back to mainland Europe to support the war effort more directly. He did ask his commander twice in a single day, though the timing was different than depicted here. More details will be discussed later. [Ref 6, 10, 11]

Chapter 6

Onward to Victory

Three days came and went, and the anxiety of Rupert's uncertain future was replaced with an excitement that could have only come from naivety. The harsh realities of the warfront in France were nothing to be excited about. Rupert, along with the other pilot and two navigators decided the best way to prepare was to spend their time studying. They studied all kinds of things; known German flight tactics and procedures, maps of the French and Belgian countryside, current European politics, and any other seemingly relevant information they could get their hands on.

With every day that passed the anticipation grew. Unfortunately for the aviators, as the anticipation increased, time seemed to slow down. Another three days later the studying finally ran dry, and they were exhausted from packing, studying, and keeping up with their patrols. Now that it was the night before their departure it seemed like the week had passed by in no time. No one could sleep so they sat in the briefing room talking about whatever came to mind.

"They call him the Red Baron." William said. "They say he's the greatest pilot in the world."

"We'll see how great he is when he flies against us!" The other pilot said eagerly.

Rupert got an uneasy feeling deep in his stomach. As their leader, he wanted to instill confidence, so he mustered his strength and said, "He'll be no match for us once we get our new planes!" Rupert was beginning to realize going to war comes with many emotional ups and downs.

The other pilot opened the door, and the warm night air blew in. He stepped outside into the dark camp. Before the door swung closed, sirens began wailing loudly from every direction. The pitch-black night quickly lit up as buildings around Sutton's Farm burst into excitement.

From the barracks to the briefing rooms, shouting echoed from every direction. The entire cantonment area was bustling. A voice finally came over the speakers to make an announcement. It was the distinct nasally voice of their commander.

"German bombing raid over London. Pilots only, mobilize and engage the enemy." The squadron commander repeated these short directions three times. They had flown to London hundreds of times, both during the day and at night, so that simple order was all anyone needed to complete their mission.

"Why can't I come?" William yelled as they sprinted to the plane.

"Too much weight now that we have the machine gun!" Rupert yelled back.

"But we have the machine gun on our patrols!"

"Yes, but those are patrols, we need to conserve all the fuel we can for the real thing!" Rupert reminded him as he jumped into the cockpit alone and screamed, "ONWARD TO VICTORY!"

"Godspeed Rupert," William said, but the engine roared to life and Rupert didn't hear him.

Once Rupert was in his cockpit, he was as sharp as a talon on one of the ospreys he admired when he was younger. As a child, he often sat by the sea and watched the great hunting birds dive to the ocean, poised to pluck a fish from the water. Now he was the one flying. He pulled up on his controls and brought his plane high into the cold but familiar, black sky. He circled overhead and waited for the rest of his Flight to take off from the crowded runway.

After circling the camp, he turned on his flashlight, signaling his wingmen to join him in aerial formation. As the fourth plane settled in on his left wing, he flashed his light at the other pilots before swinging it in the direction of London; and they were off. Within 10 minutes they saw the glow on the horizon as they approached the familiar lights of the city. Tonight, the glow was different. Giant beams of light were swinging back and forth, desperately searching the sky for the raiding airships.

Meanwhile, inside the zeppelins, Germans leaned out the windows dropping explosives. The streetlights in London looked like stars in the night sky. Factories were indistinguishable from houses, so the beasts bombed the city indiscriminately. The people of London had no choice but to endure the terror.

As Rupert eagerly cruised towards London, he thought about the last time he found himself face to face with a zeppelin, armed only with a pistol. Since then, the whole squadron modified their equipment and tactics. Now every plane was equipped with a machine gun and special ammunition.

Theoretically, exploding bullets would rip holes in the blimps, allowing the highly flammable helium gas to pour out. The pilot would then switch out the exploding bullets for incendiary bullets which would burn as they flew, igniting the gas, bringing the massive zeppelin down in a dramatic ball of flames. It sounded like a good plan, but no one had ever tried it. [FACT 15]

Rupert's heart started to race, and his stomach became queasy. The air was cold, but his nerves got the better of him and sweat droplets formed under his leather helmet. His first encounter with a zeppelin haunted his mind. He could feel the throbbing pain in his arm as if it was a fresh wound. He subconsciously tightened his grip on the controls. The memory of the waves hitting the rocks was burned in his mind and the thought of his plane crashing into the coast filled him with dread.

It's been well over a year since I last saw a zeppelin, Rupert thought to himself. I have a machine gun now but what if they changed their tactics? What if their ships are different somehow?

"No time to ponder that now, I'm in the thick of it." He whispered softly, "Onward to victory." Rupert quietly repeated this phrase like a battle cry. He had never said it before tonight and wasn't sure why he was saying it now but somehow it gave him confidence.

In the limited light available, Rupert began recognizing individual buildings and he knew they were getting close. He felt like he was one step closer to his destiny, though he wasn't sure what that destiny was. The sky above London was eerily quiet. The enemy was out there but the searchlights were passing back and forth desperate for any glimpse of the airships terrorizing the city, but they found only darkness. It was so dark Rupert was barely able to make out the pilots on his wings.

If those spotlights don't light up a zeppelin soon, we might crash directly into one, he worried.

They could see small bursts of light on the ground as the bombs exploded in various parts of the city below. Suddenly a different flash of light reflected between the clouds from somewhere in front of Rupert's formation. A second searchlight quickly passed through the same spot in the sky, illuminating the tail of one of the flying machines before it vanished like a ghost into clouds and darkness.

Rupert's heart began to race but then that odd, almost peaceful, clarity he longed for, took control of his mind. It was like all his senses were heightened. He could see and hear better. He could make decisions without having to think about anything. His plane truly became an extension of himself. If he thought of a maneuver, somehow his plane would just do it. He was as sharp as a razor, focused on his mission, nothing else in the world mattered. It was as if all his worries about crashing or being shot never existed.

He waved his flashlight at the other pilots who signaled back with theirs. In unison they all veered slightly right and gained altitude so they could swoop down on the German blimp. Rupert looked to his left to make sure the other two pilots followed his maneuver.

He scoured the darkness searching for his target but found nothing. He looked down and realized the city lights were directly underneath him. The spotlights swung wildly through the air, cutting the darkness like a knife through butter. He could hear the wailing engine of one of the blimps floating somewhere in the darkness over the city, still dropping bombs on anything that happened to be below.

BOOM! A flash of light and a loud concussion overwhelmed Rupert and his formation of small planes. He felt like a quail being hunted in a bush. A friendly anti-aircraft shell exploded about 500 yards from his squadron.

The artillerymen must have gotten a glimpse of the zeppelin's tail as well, Rupert thought, looking around. Two of Rupert's wingmen veered off to the left and disappeared into the darkness. Just then the light swung back and paused on the bottom of Rupert's plane.

For a moment Rupert was illuminated, caught like a deer in headlights until the beam of light swung in front of him, like a mystical light guiding him to his destiny.

Rupert found himself in a moment of pure darkness, waiting for his eyes to adjust. I hope the anti-aircraft gunners recognized friendly planes in the area, he thought, but that is out of my control.

Rupert and his one remaining wingman scrambled forward into the darkness. The searchlight found its mark on the silvery airship that lit up like a heap of metal in a junk yard floating through the sky. Rupert estimated he was less than 400 yards from the monstrosity. He grabbed the handle of his machine gun, but another concussion of friendly anti-aircraft caused him to lose his grip on the plane's controls. He immediately grabbed them with both hands and pulled back with all his might. He maneuvered his aircraft up and to the right. He completed a large circle, lining himself up to get his first shot at the beast.

In the chaos, the other pilot steered downward and was nowhere to be seen, leaving Rupert as the lone pilot against a zeppelin in the fight to save London. He stared down the barrel of his machine gun, now pointed directly at the blimp. The beam from a searchlight created spires of light towering into the sky around the zeppelin. Shadows danced on the grey canvas as each of the blimp's panels fluttered independently, like flags in a storm.

He squeezed the trigger, and watched bullets shoot out, igniting as they flew. They glowed through the darkness of night, spouting a tiny flame along the entire trajectory before disappearing into the night. The handle of the gun was freezing cold, but Rupert didn't notice. Another anti-aircraft shell burst a few hundred yards away, but Rupert was mesmerized by his glowing bullets that looked like a string of tiny shooting stars.

He flew alongside the silvery giant, firing upon it until either it or the spotlight turned, leaving his target camouflaged in darkness once again. I'll circle back for another pass, Rupert thought. He reached down with one hand and removed the empty drum of ammunition from the machine gun. As the drum dropped to the floor of the cockpit Rupert realized something.

The bullets that glowed in the darkness were supposed to ignite the helium gas escaping from the holes, but he never fired the exploding bullets to rip the holes in the first place.

"I JUST WASTED AN ENITRE DRUM!" He screamed out in frustration. He was left with two drums of hole-rippers and one drum of bullets to spark the flame.

As the spotlights searched between the clouds for his target, he gave his best guess where the zeppelin should be and oriented his plane. He gripped the handle of the machine gun again just as the spotlight lit up his plane again. The sound of bullets immediately whizzed past him forcing him to duck down. He peeked up over the safety of his cockpit, just enough to see the small flashes of light from a single point in the darkness surrounding him. Now, I've got them, he thought.

The German's bullets ripped past Rupert; they sounded like supersonic bees, buzzing by until one hit its mark. A loud crack of fracturing wood came from somewhere on Rupert's plane, but he continued forward without hesitation. He pulled the trigger firing half the drum, hoping to rip holes in the blimp. He closed in and the spotlight finally found its target just in time for Rupert to see a German soldier hanging out a window, dropping some sort of bomb on the city below.

A newfound sense of urgency inspired Rupert as he imagined what that bomb might do to innocent people below. He turned downward to avoid the blimp then quickly gained altitude to circle around for another pass. He wanted to attack from above to maximize the chances of ripping holes in the canvas.

Onward to Victory

He was lined up for his attack. The spotlight's beam illuminated the blimp from below making it look like a runway. From around the balloon's edges, beams of light escaped into the heavens which created shadows dancing eerily like ghosts in the night. The searchlights stayed locked on target allowing Rupert to strafe the entire length of the mammoth balloon.

He turned for another pass while loading the ammunition designed to ignite the explosion. I have to use this sparingly, he thought. He lined up his attack directly at the zeppelin. He fired down upon it just like his last run. Each bullet glowed brightly in the black night. Time seemed to slow down; it felt like an eternity passed as he held the trigger.

Nothing happened. What if I run out of ammunition, he worried. If I didn't rip any holes, then I'm wasting my last incendiary bullets. He realized he needed to pull up for another pass because if he scored a hit this close, he would surely be swallowed by the explosion.

He instantly let go of the trigger. One last glowing bullet exploded out of the barrel, rapidly spinning towards the zeppelin. He steered up and to the left as the bullet burned through the night sky like a shooting star that was good for one last wish. The zeppelin was now directly beneath Rupert's plane as the fiery bullet found a pocket of helium gas escaping from the blimp. The flaming bullet ignited the gas just as they hoped.

A giant fireball erupted under Rupert and the flames whipped out, singeing the underside of his plane. It felt like fire kissed the back of his neck before an intense heat surrounded him as if he jumped into the pits of hell. He instinctively covered his face with his scarf as he flew out of the explosion and into the night that now felt colder than ever. If he hadn't started to pull up when he did, he would have been engulfed in flames with the zeppelin.

The Londoners watching from the ground began cheering as what was left of the zeppelin fell to Earth like a flaming meteor. Rupert realized he chased the blimp outside the city, and it was falling into an empty field. It could not have gone more perfectly. He triumphed with a contentment he had never experienced before. It was greater than being captain of his football club, greater than being named head boy of his school house, it was greater than anything he ever imagined. He continued to hold his turn in a large circle around the spot in the air the zeppelin had just occupied. Two searchlights found Rupert and followed him around his victory lap, as if trying to celebrate this anonymous hero.

As the flaming zeppelin fell, Rupert was dancing on the clouds. It crashed to the ground and a devilish sound of metal twisting against its will overwhelmed the silence of the night. Flames leapt into the night sky as if a portal to hell opened and sucked the zeppelin in. Rupert straightened his controls coming out of the turn and set a course for the airbase at Sutton's Farm. The spotlights below followed Rupert and the Londoners watched their hero fly off into the night.

The worries he had about dying vanished and that odd clarity he only felt in battle was replaced by pure bliss. He was on top of the world, elated; there was nothing he couldn't do. He scanned the skies for his wingmen hoping they would be illuminated by one of the searchlights still scanning the darkness. He didn't spot anyone in the black night but what the lights did find made Rupert sick to his stomach.

Off in the distance behind him the spotlights converged on another zeppelin. Rupert's triumph was short lived. His bliss was gone, replaced by another feeling but it wasn't that peculiar calmness; it was fear.

He took a deep breath, checking his petrol levels and the other gauges. Everything seemed to be in working order and he had enough fuel for one last skirmish.

He began a large turn and almost collided with two planes from his sister squadron. He followed them back over London toward the second blimp. He switched to the fresh drum of hole-ripping bullets. That, and less than half a drum of the incendiary rounds was all he had left.

Five spotlights from every corner of the city below now converged on the remaining zeppelin as a single light continued to search the skies for any other flying beasts hiding in the dark.

BOOM! BOOM! A double burst of anti-aircraft artillery shook the tiny planes in the sky from the opposite side of the zeppelin. They must have seen us all fly away after the first zeppelin fell, Rupert reasoned.

The squadron ahead of Rupert immediately dumped their ammunition, firing it wildly at the zeppelin but pulled up before they were close enough to be effective.

"YOU'RE TOO FAR AWAY!" Rupert yelled aloud realizing he needed to lead the charge. He got so close he could see silhouettes of individual soldiers running in the zeppelin's cabin, furiously trying to drop their stockpile of bombs while another soldier steered the beast away from the city in a hasty retreat.

Rupert approached from the rear of the blimp. He lined up his shot and fired upon the zeppelin, scoring hits all down its left side, before pulling around for another pass. He loaded the half drum of incendiary rounds. He lined up his shot, now attacking the blimp from the front, aiming at the exact line where he emptied the first drum.

BOOM! Another explosion from somewhere behind him shook Rupert. A ringing overwhelmed his ears and he quickly realized he had the worst headache of his life. He was in pain, but he decided to complete his mission even if it killed him. The anti-aircraft burst caused him to miss his opportunity so he swung around for another attempt. This time Rupert passed underneath the zeppelin and through the light shining up from below. He was temporarily blinded but hopefully the men operating the spotlights would inform the artillerymen below that friendly aircraft were in the vicinity.

BOOM! Another explosion, this time from above the zeppelin he was underneath. He flew back around and lined himself up for his one last pass. The lights and explosions had nearly blinded him. He squinted his eyes trying to see anything and made out what appeared to be a large swath of canvas.

If I didn't rip any holes in the balloon, then that artillery surely did!

Rupert pulled his trigger. The first incendiary bullet ignited inside the chamber of the machine gun. It spiraled down the barrel until it exploded out of the gun, burning a trail of light into the black night. The first bullet was followed by a hundred more and from the ground it looked like a stream of fire flowing from the lone plane like the wrath of God as David fought Goliath.

The artillery ceased. Rupert held steady on his controls and the trigger of his gun. He poured all his might into his mission.

The machine gun clicked, and Rupert recognized he was out of ammunition. It's in God's hands now, he thought with a sigh as he watched the bullets glow through the night above his intended target. I missed, he thought, as he began to turn. But the last three flaming bullets found a small pocket of helium gas that had risen far above the zeppelin. It ignited creating a tiny explosion above the blimp which burned off in an instant, but it did something else.

Just before that small pocket of gas fizzled out, it met a stream of helium pouring out from the top of the balloon. Fire leapt from the pocket of helium to the stream and for the second time that night, a giant ball of fire lit up the night sky over London.

Rupert flew around the city; people poured into the streets waving their hats and cheering from below. I did it, he thought in disbelief, I really did it.

Back at Sutton's Farm he came in for a hard landing; something on his landing gear must have been damaged. He taxied as far as he could, barely making it off the runway. Men from his flight crew, as well as the other pilots in the squadron crowded around Rupert's damaged plane cheering as if he scored the winning goal in the last seconds of a world championship.

Rupert hopped down into the crowd, and they hoisted him onto their shoulders.

William cut through the crowd and said, "They've just radioed us, you did it didn't you? You shot down the zeppelin?"

Rupert sat upright, balancing on people's shoulders, and held up two fingers.

"Two?" William had no idea what Rupert meant and then it hit him. "You shot down two?! HE SHOT DOWN TWO BABY KILLERS! LEFTENANT ROBINSON IS THE HERO OF LONDON!" William screamed from the middle of the commotion.

The three other planes in his Flight had all been hit; one by the first anti-aircraft shell, the second by machine gun fire from the zeppelin and the last by the anti-aircraft explosion Rupert heard off in the distance as he was closing in on the first zeppelin. The pilots from the other squadrons were all scared off by the anti-aircraft artillery. An examination of Rupert's plane revealed, it too, had been torn to shreds by the machine gun fire and shrapnel.

The next morning Rupert awoke in the bunk he had called home for nearly two years. He should have been reveling in his newfound glory as the Hero of London, but instead he just laid on his cot with all his thoughts on France and the perils he would face there. The zeppelins had very little means to maneuver against his agile biplane but the German pilots over France would not be bested so easily.

He would never know, at that moment, as he solemnly packed his bags to depart England, Londoners were toasting his name in pubs across the city. On the other side of the airstrip, the sun peaked over the trees at the edge of the woods casting long shadows on the morning dew. Rupert, William, and the others boarded a bus, bound for a ship, which would deliver them across the English Channel to the treacherous front lines of Europe's Great War. [FACT 16]

FACTS

15.   As described in the narrative, the planes were outfitted with two types of bullets. One type would explode, ripping large holes in the canvas and the other bullets were incendiary rounds which would ignite the flammable gas, bringing down the whole zeppelin in a dramatic and dazzling ball of flames. The incendiary rounds gave the English the idea for a non-incendiary design which is still used today, commonly referred to as tracer rounds. [Ref 10, 11]

16.   In the early morning hours of September 3, 1916, it was Lieutenant Leefe Robinson who was dubbed the 'Hero of London.' He fired the shots that brought down the first zeppelin destroyed in battle. Just as they theorized, the special bullets ignited the zeppelin in a giant ball of fire which fell to the ground. Londoners cheered as it fell. Thousands of people came to visit the field behind the Plough Inn in a small village called Cuffley, where

the remains of the zeppelin landed. Many visitors took souvenirs from the burnt heap of scraps.

There was only one zeppelin that night. Once the British developed the exploding and incendiary bullets, and Leefe proved the concept, zeppelins became obsolete. Two thirds of all the German zeppelins were subsequently destroyed or disabled. [Ref 6, 10, 11]

Leefe did not report to the warfront the following morning, as depicted here. He shot down the zeppelin on 3 September 1916 and didn't leave until March of 1917 where he was assigned as the Commander of the 48th Squadron.

Chapter 7

The Somme Offensive

October 1916

"To the first man to bring down a zeppelin; three cheers for Leftenant Robinson!" William screamed in a crowded French pub. The building was made from old cobble stones stacked around large timber beams all held together with some type of cement plaster. It was a quaint, dimly lit bar just a few minutes' walk from their new barracks in Northwest France. There were a few locals, but it was mostly filled with pilots and crew all dressed in the best uniforms they had with them.

Beer sloshed and spilled every which way as the crowd raised their pints and shouted, "Hip, hip, hooray!"

"Speech!" yelled a voice from the crowd.

Rupert was begrudgingly hoisted onto a barstool. A small red medal pinned on his chest stood out against the dark green olive-drab uniform. Earlier that day the commanding officer of his new unit officially recognized Rupert as the first person to shoot down a zeppelin. He was awarded the Victoria Cross, the highest honor for gallantry in the face of the enemies of Great Britain. [FACT 17]

I should have prepared something, Rupert thought. He hastily came up with something to say, "They wouldn't dare meet us in broad daylight, so they sent their warships in the night. Now they can send their airships all they

want! We will meet them in England, we will meet in France and Belgium
and soon we will meet them in Germany, and everywhere we meet them, we'll
turn them all to ash!"

"Hear, hear!" The crowd replied over the sound of clinking mugs
before the bar went back to its normal rumble of conversations.

Rupert, William, and the other airmen from the 39th Home Defense
Squadron settled into their new living accommodations which consisted of a
makeshift barracks that looked, and smelled, as though it had been converted
from a pigsty. They left their planes behind and had to get used to the new
Bristol F2s. They were the newest model which had been sent directly to the
front lines. They were equipped with state-of-the-art engines and had two
proper seats so the pilot could focus on flying, aided by their observers who
would navigate and fire the machine gun which was mounted behind the
second seat.

When they weren't in the air, they received briefings on various topics
like German aerial tactics, weather patterns, or Allied efforts. The worst
briefings were the statistics. At that time the average life expectancy for a
pilot on the western front was 11 days, which equated to just 20 hours of
flight time. [FACT 18]

Days… was that really all we have left on this Earth? Rupert thought,
lying awake one night.

Rupert was now the official commander of his Flight. The 48th
Squadron was designed to have 18 planes divided evenly into 3 Flights. Due
to the war the squadron only had 12 planes. Each Flight also included the
entire ground crew required to support the pilots and planes. He led with
pride, though at times his confidence faltered. The men thought they
volunteered to relieve pilots who would rotate back to the rear for a break.
The cold hard truth was they were sent to France to backfill crews who had
all been killed in action. This sobering reality was often the cause of heavy
drinking.

He was in new surroundings, but Rupert was still haunted by a feeling
of impending doom. He thought it would subside as he settled in, but it

festered like an untreated wound. In his darkest moments his mind dwelled on his own demise or worse. He was tortured by the understanding that he dragged his men to this certain doom alongside him. They were good men who deserved better than this fate.

Perhaps worst of all was the waiting. Rupert and the aviators from England were in a probationary status for their first two weeks in-country. They were not authorized to fly in combat, rather they would fly once every two days, well behind the front lines to get acclimated to the plane and new environment. This created lots of down time which was filled with nothing but uncertainty, angst, and alcohol.

He wrote to his mates back at Sutton's Farm telling them of the French countryside and the beauty they encountered on their journey, but he also wrote about the alarming truth; he and the others from the 39th would probably never return home. As a leader, Rupert ensured the members of his Flight were as prepared for their first mission as they could be, but that was all he could do.

On the morning after their probationary period ended, Rupert laid motionless in his bunk, only half awake. His eyes remained closed even though he could tell it was getting light out. He was probably going to miss breakfast and the morning formation, but he didn't care. He took a deep breath and allowed his mind to be clear, focusing only on the present moment. The sun was climbing over the tops of the trees surrounding the barracks. The runways began to soak up the morning light as the dew in the surrounding fields glistened.

In an instant, all the peace Rupert was clinging to disappeared. His barracks shuddered, and his eyes shot open. A loud explosion followed by another, sent tremors through the camp and adrenaline through his veins but Rupert continued lying flat on his cot.

The Germans must be attacking, he thought without any emotion. Rupert was frozen in what little comfort was lingering, either unable or unwilling to move at all. The artillery is probably meant for the trenches and those are at least a mile away, he figured. It's time to face my destiny. Today I fly.

Whistles echoed like screaming banshees through the forest which separated the air strip from the front lines. Machine gun fire cracked from somewhere far away.

Rupert was about to find out the French artillery moved overnight. There were at least a dozen guns directly across the field, in the same clearing as his barracks. The black tubes of the great cannons were pointed up towards the front lines as men hustled around each one, yelling and loading the guns. They fired their first volley with an explosion that sent a massive shockwave through the clearing. The concussion wave hit Rupert's body.

He sprang from his bed, quickly dressed and instinctively ran to his plane, alarms rang in every direction, as planes took off, one after the next. Rupert got to his, but it hadn't been prepared to fly. His crew was nowhere to be seen. He headed to the briefing room hoping to find a familiar face.

Much to his surprise, he walked into a packed room where a briefing was already in progress. He looked around at the group of French and English aviators listening to a French Colonel.

"Now is no time to dilly dally Lieutenant!" The Colonel spat angrily at Rupert's tardiness.

He found William who quickly caught him up to speed.

"The only planes flying now are reconnaissance." He whispered, "This Colonel is trying to brief us on the counterattack we are supposed to help with, but it sounds like they are making it up on the spot."

The commander had been briefing in French then switching to English but now was speaking only in French. He stopped briefing at certain points to ask pointed questions to his staff officers. Two of them had a heated exchange before it burst into yelling.

Another officer stopped the argument before the Colonel started briefing again.

"They *are* making it up on the spot!" Rupert whispered to William in disbelief.

Finally, it appeared that a plan had been hatched and the French crews began leaving the room.

"Sir, what are the commander's orders?" Rupert stopped a French Major just before he walked out.

The Major stayed behind to brief the English pilots. The whole plan was dependent on the infantry repelling the Germans who, at that moment, were advancing across no-man's land to attack the Allied trenches. The plan was a massive counterattack in which almost every pilot they had would be in flight, each with a different mission. Some would be reconnoitering the battle, or range finding for artillery, most would provide bombing and machine gun support directly to the infantry.

The Major explained that the Englishmen had a special mission. They would be going deeper into enemy territory than anyone else. Their mission was paramount; find and destroy the German artillery. It was a perilous task, but if successful, could turn the tide of the battle, which in turn, could win the campaign.

"This counterattack sounds like they want us to do the exact same thing the Germans are doing!" William said to Rupert as they walked across the field toward their barracks. "If it doesn't work for the Germans, why would it work for us?"

"Based on what that Major said, I believe you and I are to be the deciding factor. The Germans are only attacking on the ground while we will attack from the skies." Rupert reasoned. "What worries me is if we are going further than the other planes and the Germans dispatch their fighters, we will be the first ones they reach."

"Why can't the Frenchies use their artillery to attack the German guns? Why send us?" William complained.

"They are out of range." Rupert guessed. "If the French move their guns any closer, they could be taken out by the German artillery before they are set up and ready to fire. I wish I had understood what the Colonel was briefing, I would have requested a Flight of fighters to escort us!" Rupert felt like he let his men down by not challenging the plan before the briefing adjourned but since most of it was spoken in another language there was little he could have done.

The counterattack was to begin an hour before dusk when the Allied forces would have the sun at their backs. Then, as night fell, the artillery and Flying Corps would send up flares with parachutes to keep the battle going into the night.

By the time the briefing ended it was already lunchtime, but Rupert and William had no appetites. They spent the early afternoon lying on their cots trying to mentally prepare for what was ahead. Unfortunately for Rupert, memories of his past encounters with the zeppelins dominated his thoughts.

He flinched on his cot, hearing bullets whiz past his head, though this time it was just his mind playing tricks on him. He could hear the bullets ripping the canvas and splintering the wooden structure of his plane. His only comfort was knowing that when the bullets were really flying, he would get that peculiar feeling of extreme focus. When he was flying his plane through dire situations, he never worried about anything, he just focused on what needed to be done and did it.

It was a very odd thought; after all he had been through, the only thing to relieve his fear of war was the idea of being back in combat as if he was an addict having withdrawals.

The battle in the trenches raged on into the late afternoon. Artillery barraged the Allied lines but mostly just shook the soldiers who were protected by the mud walls of their deep ditches. The French artillery, on the other hand, was firing non-stop into no-man's-land causing immense physical and psychological distress to the attacking Germans. For hours, wave after wave of German infantry hit the Allied line like a relentless ocean. Finally, the last wave of Germans attacked as if the tide was going out. The German's regrouped in the same trenches they left hours earlier. What was the point of it all?

The hour of the counterattack drew nigh, and the flight crews prepared the planes. A mile away from the air strip, the infantrymen in the trenches had no emotional support so they leaned against their mud walls for any encouragement they could find. They received their orders and prepared for what felt like a suicide mission.

The tired soldiers sharpened their bayonets, cleaned their rifles, loaded ammunition, and tightened their boots. Just in case, they checked their gas masks, prepared to pull them out at a moment's notice. No one in the trenches looked anyone directly in the eyes. Making eye contact in those dire moments would have been like staring into a mirror of their own soul with the understanding that they were probably living their last moments on Earth.

The Somme Offensive

Rupert looked to the sun that was obscured by a dusty haze. He watched it slowly sink to Earth as the late afternoon gave way to an evening sky. He knew the moment was upon him. He and William climbed into their cockpit without speaking a word. All the activity on the runway made it seem like every pilot in France was taking to the sky, linking in formations overhead. Whatever personal comfort Rupert clung to was left on the ground in his barracks in the French woods.

The sky filled with planes like a murder of crows chaotically swooping every which way. Through the woods, the ghostly sound of whistles echoed from the trenches signifying the start of the counterattack. As the whistles screamed like banshees below, every plane in the sky turned away from the sun like a flock of birds scared by a loud noise.

The planes organized into attack formations, some dropped in altitude and flew just above the trees while others cruised high above the rest. Rupert led the squadron towards their first major battle. He tried to calm his nerves, but his heart was pounding in his head. He settled into his designated position in the aerial charge; above, and about a quarter mile behind the first planes.

In the fields below, the infantry climbed out of their trenches and charged across no-man's-land to certain death. They heard the low rumble of planes buzzing behind them which gave them just enough confidence to push forward. Whistles continued to blare as men screamed up and down the line. The last of them hopped out of the trench's relative safety just as the first planes flew low over their heads, leading the attack.

In the German trenches across no-man's-land soldiers looked out to the desolate terrain they had just retreated from. The burn of the evening sun shone through the dusty yellow haze, obscuring their view. Whistles echoed from across the charred and battered field as the rumble of plane engines and the stampede of infantry began to shake the earthen walls of the German trenches. Sergeants barked orders to load weapons and assume fighting positions, their hearts pounded in anticipation. The Germans didn't think the Allies would launch a counterattack this late in the day since neither side had an advantage in the dark. Attacking was always a disadvantage compared to being in fortified defensive positions.

German soldiers quickly loaded their rifles while others carried long strands of bullets that were all linked together like giant necklaces. They fed one end of the strand into the machine gun trays before slamming them closed, locking the bullets in place, now ready to fire. That distinct metallic crunch echoed up and down the defensive line before an uneasy silence fell over them. They had barely survived their attack that day, now they were facing a counterattack.

In hopeless angst, they squinted across no-man's-land directly into the bright, dusty light. Silhouettes began to appear on the horizon like mirages in the desert. They aimed, looking for the Allied attack, but couldn't differentiate men from shadows. Black specks in the air above slowly grew larger like a swarm of locusts descending upon them.

From high above, and with the sun behind him, Rupert watched hundreds of troops charging across the war-torn ground, running towards the German line. The land had already been shelled many times and was left barren with only the occasional broken tree still hanging on. The ground was stained with blood. Thousands, if not tens of thousands, of men from both sides had already perished in those fields in West Flanders. Rupert looked ahead to the German lines just in time to see the machine guns begin firing their first shots. French and English soldiers charging through no-man's-land began falling.

"Weather seems nicer up here!" William yelled to Rupert over the wind in their cockpit.

"Let's hope it stays that way!" He yelled back.

"We are approximately one minute from the first trench!" William called out. "It looks like two more trenches past that one, spaced at 50 yards apart!" The German trenches looked like a maze you might find a lab rat trapped in.

One bold German machine gunner pointed his barrel up and fired it wildly in the air. Rupert watched from above as an Allied plane crossed over the first German trench. The observer dropped grenades and fired at the Germans below, but the machine gun found its mark. Fire burst from the front of the plane before it exploded. What was left of it crashed just behind the first trench.

That was the first pilot to die today, Rupert thought to himself. The queasy feeling deep in his gut told him it would not be the last.

From somewhere beyond the trenches, in the thicket of trees, a burst of bright light exploded into the evening sky.

"Artillery!" William screamed. "One o'clock, one mile, one minute!"

Rupert looked back at the advancing troops, a huge explosion in no-man's-land flung soldiers into the air along with mud and rocks. He felt like he was in some horrible dream that he couldn't wake up from. How could this be real? He thought to himself.

Rupert snapped back to reality. He wasn't watching a horrible dream; he was living one.

"We must accomplish our mission!" Rupert yelled to William who was signaling the target to the other observers.

The other planes in the squadron broke off for their respective missions while Rupert's Flight of four planes continued straight over the first enemy trench. William watched the other planes form a circle in midair; one plane at a time swooped in to fire along the German line while the others circled around for another pass. This way one of the planes was always firing on the enemy, never giving them a break. Rupert watched long enough to see French and English planes begin to explode or catch fire, one by one crashing to the ground.

He winced at the sound of stray bullets whizzing past as they flew by the second trench. "Why aren't they firing anti-aircraft at us?" Rupert yelled.

"I don't know but if they do, be sure to fly around it!" William said.

"Oh, wonderful William, I hadn't considered that!"

The few Germans remaining in the second and third trenches ran to their machine guns. Rupert could fly through potshots from individual soldiers, but machine guns were a much different threat. He pulled back on his controls leading his Flight to a higher and hopefully safer altitude while they lined up their attack. William turned, looking back toward the first trench. Another plane below got hit by direct machine gun fire and burst into flames. For a moment, it continued flying as if it would land, then it turned straight down and crashed in the trench, crumpling into a smoldering wreck.

Rupert noticed this as well, and the all-too-familiar feeling of impending doom filled his mind. Come on, he silently encouraged himself.

We must get to that artillery. You are leading these men after all; they are counting on you!

"20 seconds!" William yelled, pointing at the ground ahead of them. Rupert identified their target.

With his wingmen close behind, Rupert signaled to ready the machine guns. William took this opportunity to prepare a hand grenade by pulling out the pin. As long as I hold this little lever, the spoon they call it, William thought to himself, then the fuse won't ignite, and the grenade will never explode. After the spoon detaches then I have three to five seconds before it goes off.

Machine gun fire from below cut through the warm evening air and began to walk its way through Rupert's small formation of planes. He signaled for his Flight to spread out while maintaining the higher altitude. He would lead a diving attack almost directly on top of the target, like a hawk diving at its prey.

"Ten seconds!" William yelled to Rupert over the noise and chaos around them as he pointed forward at their intended targets. William couldn't see down over the side of the plane, so he adjusted the way he was sitting. At the same time, Rupert deliberately began a quick decent towards the German artillery but hit a patch of turbulence and the combination shook the plane hard.

This dislodged William's hand as he was adjusting in his seat. His hand slipped off the bench and the grenade he was holding fell. It hit the floor of their cockpit with a thud and the spring-loaded-spoon popped off, causing the internal mechanisms to ignite the fuse.

"OH GOD!" William screamed.

"What?" Rupert yelled as he pointed the plane down so sharply it felt like they must be pointed straight at the earth.

The g-forces held William's back to the seat, but he lunged forward, trying to grab the grenade that was quickly cooking off on the floor. Rupert leveled the plane and William was thrown into the seat in front of him, as the live grenade rolled between his feet. He wrestled out of his harness and dove into the tiny space below his seat.

"What's going on back there?" Rupert said, "There is a machine gun nest you could hit right now!"

The Somme Offensive

As Rupert said those words the machine gun started firing. William's fingertips touched the grenade, and he snatched it up as quickly as he could.

This thing is going to blow! He thought desperately as he aimlessly flopped the grenade over the side of the plane and recoiled as if he had just touched something repulsive. He was curled up, sunken as deep in his seat as possible, waiting for the explosion. He held his breath and counted. There was a psychological weight crushing him but for every moment the grenade didn't explode it lifted just a little.

The machine gunner aimed directly at the plane as they flew over top of him, but his aim was short. The tracer rounds glowed through the air, curving through the sky as he adjusted his aim. One more second and his stream of bullets would surely cut their plane in half. William's desperately flung grenade fell to earth and exploded less than two seconds after he abandoned it. They were just out of the grenade's effective range and William let out a huge sigh of relief. Coincidently the grenade exploded directly above the machine gun nest that was firing at them. It immediately fell silent, and the line of bullets that had been crisscrossing around them ceased just before intersecting with their plane.

"Great hit William!" Rupert commended, "You lined that up perfectly!"

"Oh… it was nothing." William said through a weak laugh.

"Ha ha! Brilliant hit!" Rupert exclaimed again.

William laughed in disbelief before regaining his confidence, "Now on to the next one." He laughed.

Rupert finally found that calm focus he longed for. All the normal worries of life were gone, his only thought now was to deliver his Flight to the target. They were seconds away. They were so close to the ground William thought he heard some German soldiers yelling so he gave a cheeky wave as he dropped another grenade.

They passed the last trench and saw the artillery in a clearing poking out of the woods like straws sticking out of a drink. There were four big guns, spread out in a line. Two guns were to their left and two were to their right. Rupert veered toward the two on the right. Holding the plane steady, he noticed some motion in the woods past the guns. The motion stopped and was replaced by flashes of light.

Bullets whizzed past but Rupert stayed focused on the artillery. His thoughts were perfectly clear, his fear was gone. He was in the moment, focused on his mission, dodging direct machine gun fire and sporadic shots from soldiers below, though in an instant all that would change.

The machine gun fire he had successfully evaded traversed left, then right, through the formation, peppering two of the other planes as all four swerved to avoid being hit. Another second went by, and the machine gun landed a direct hit on the third plane in formation. William quickly turned to look at it. Its propeller was badly damaged. The plane will likely crash, he thought, feeling hopeless in the moment. He wanted to help but was powerless. He knew he had to focus on his mission. They're on their own for now, he thought in despair as he turned forward to find his target.

The machine guns continued to repel the predators in the sky. The plane that just got hit burst into flames, which caught Rupert's eye. The next thing he saw would haunt Rupert for the rest of his life. The pilot of that burning plane was shot in the chest. The flaming plane, now without a pilot, did a nosedive. It crashed into the ground below which was now covered by long shadows from the setting sun. He was dead before his plane hit the ground, Rupert thought, not sure if that was better or worse.

Rupert was petrified by remorse. He just witnessed one of his direct subordinates get killed in action because he was following Rupert's orders. Rupert's peculiar calm focus was replaced by new disparaging thoughts. That pilot had not just been a subordinate; he had been a friend, and now he was gone.

A stray bullet ripped through the canvas of a wing about eight inches from Rupert's head which brought him back to reality. The three remaining planes were finally able to engage their target, now in range.

The artillery was defended from a ground assault by trees and mounds of earth but from the skies they were like fish swimming at the top of the sea, exposed to Rupert who felt like a hunter ready for his kill. The planes flew even lower to ensure the most accurate attack possible. The rear-man in each plane threw grenades and opened fire directly at the artillery as the pilots evaded enemy machine guns.

The grenades fell to earth and exploded as the planes sped past their targets, not seeing if they hit anything. The three planes pulled back looping around for another pass, before swooping back down.

BOOM! The battery of enemy artillery erupted into the evening sky, sending out a shockwave that transmitted tremors of fear through the aviators in the sky. Rupert found courage by renewing his resolve to complete his mission after seeing the artillery open fire at the approaching infantry. Each volley the artillery fired could be killing dozens of infantrymen. The trio of planes finished their turn, and this time approached the gun line from the side to gain enfilade fire, so they could hit all four guns in one pass. William threw three grenades while the two other gunners continued firing the heavy machine guns.

A German machine gun was now trained directly on the lead plane, which happened to be Rupert and William's. He squeezed the trigger and began firing at them, the tracers glowed in the twilight of the evening, but the rounds landed far behind Rupert's plane because the gunner was aiming where the plane was, not where the plane was going.

The machine gunner realized his rounds were way behind the plane he was aiming at but made no adjustments because he saw his tracers land directly on the last plane in formation, which crash-landed a moment later.

Did we get them? Rupert wondered as he flew past the last gun in the line. BOOM! Another volley fired into no-man's-land, answering Rupert's question. He looked to his wingmen to signal another turn and realized there was only one other plane with him. He hoped the third plane had to pull out of formation and would soon rejoin them or meet them back at their base. He wanted to scan the sky to find them but didn't have time, he had to focus on his target.

As they approached the artillery from the other side, the first gun they came to looked different. There was a hunk of debris mangled on top of it. Rupert went white in the face as he realized it was the plane now missing from his formation. He looked behind him and confirmed that he still had one wingman.

They had already lost so much; Rupert was worried it would all be for nothing, so he made every bit count. Again, they fired their machine guns and threw grenades. One of the grenades from the previous run must have hit its mark because one gun was silent, and soldiers were lying on the ground all around it, some screaming in agony.

"How much ammunition?" Rupert yelled to William to get a status report, after that pass.

"One drum left, and three grenades!" William yelled.

Rupert checked and had plenty of petrol, but his plane would not be able to take much more damage. He decided they would have to return to base soon.

"One more pass!" He yelled to William, "Give them everything you've got!"

He tried to turn hard up and to the left to circle back but his plane only went up, for some reason it wouldn't turn. He looked around and saw some small damage on his left wing, but the tail's top connection had popped out of its vertical positioning. He gained altitude and leveled off, shouting to William.

"Try to fix that rudder!"

William lunged back, hanging halfway out of the cockpit to hit the tail, trying to force it back into its proper position. The tail of the plane had been hit at least twice, but William managed to reconnect the rudder to the cable so it could function. The second plane followed, and together they circled back for one last pass at the guns.

"Just two left!" William shouted after seeing the artillery fire another volley. They swooped in, focused on their target and for some reason the machine guns below stopped firing at them. William threw his last grenades hoping to take out at least one of the two remaining guns.

As the last grenade left his hand William caught a glimpse of something behind them.

"GERMAN FIGHTERS!" he screamed, forcing Rupert to pull hard to the right.

Rupert and William's last wingman was now taking direct hits from a German fighter behind them. A moment later, William and Rupert's last wingman crashed into the ground with a lackluster thud.

Rupert was forced to abandon his assault on the artillery. He pulled up to engage the incoming fighters who now turned their total focus on the only plane left. It was getting dark, soon the pilots wouldn't be able to see each other at all.

William leaned into his machine gun and said, "Well ol' boy, it's been fun. Say a prayer and get ready to meet your maker!" Any hope he harbored throughout their mission was chased away by the fighters.

Rupert barely heard him. He was too focused on dodging the bullets that continued to pepper their plane. He whipped up to the left then turned hard to the right until he was pointed directly up. One German plane followed him as best he could while letting out a burst of bullets whenever he thought he had a clean shot. Another German plane looped back and the third had just kept flying straight. Rupert turned his plane down hard, almost going into a nosedive. The German couldn't match the maneuver and broke off. Rupert leveled off right behind one of the German fighters.

"THAT'S THE RED BARON!" William screamed happily as he recognized the pilot from a picture. Rupert turned the plane slightly to give William a clean shot and he immediately opened fire, not wanting to miss this golden opportunity. His bullets struck the tail of the Red Baron's plane, but he turned away and continued to fly.

Rupert turned to follow but the other German fighter had just finished its turn and ended up directly behind them. Unlike the first German pilot who only fired when he thought he had a shot, this pilot squeezed his trigger and didn't let up until his machine gun was empty. It forced Rupert to pull up, losing the advantage he had on the legendary ace.

"Come on mate, don't lose him!" William pleaded for another shot at the Red Baron.

Rupert kept flying, trying to get away but the three German fighters were relentless. William stopped firing back. He had fired every last bullet they had. All out of grenades and ammunition, Rupert had no choice but to abandon his mission. He needed to evade his hunters and get back to friendly lines.

The sky towards the Allied line was still glowing a pale yellow in the sunset. The sky further into enemy territory had reached the dark shade of blue just past twilight. I can use the darkness to my advantage, he thought.

"Don't worry William! I'll lose them in the darkness then double back!" He screamed as he pushed the controls forward swooping down low, before pulling up to the right, slipping into some clouds farther behind enemy lines. He pulled up even more, going totally vertical before turning hard to the left and dropping rapidly.

Rupert glanced around looking for the enemy fighters in the darkness. There is no way they kept up with those maneuvers, he thought, holding steady for a few moments. The German fighters were nowhere to be seen. Night had fully descended upon the battlefield, but flares slowly sank over the infantry in the distance. I suppose we can turn back now, he thought. He made a wide turn and settled into a high-altitude position, hoping to avoid any more contact.

"We might just make it out alive after all!" Rupert cried out into the night sky.

Not one minute later, bullets started whizzing by them and all the hope he had deflated like an untied balloon. He turned hard to the right, pulling up and gaining distance between himself and the German fighters. He looked around searching for a way out; he was desperate. I must get us home, if not for me then for William's sake, Rupert thought. After all, I'm the one who dragged him into this mess.

An explosion on the ground lit up the sky, telling Rupert he failed to destroy the artillery. His compass was damaged, so he looked around to figure out what direction he was going. His only hope was to fly back over the chaos and into friendly territory, but he wasn't sure if his damaged plane could make it, even without the threat of enemy fighters on his tail.

Unfortunately, he would never get the chance to find out. The Germans caught up to him again, so he took more evasive action, cutting hard to the right going into a nosedive. He pulled back as hard as he could, and his plane soared above some clouds into the darkness. He slowly descended through the clouds and saw no-man's-land drawing near.

"I think we did it, old boy!" He exclaimed to William. But before he heard William's response tracer rounds seemed to surround the plane yet again. The tail got hit again, then the left wing caught fire. The fire burned and dragged the plane down to the left. What little control he had left allowed Rupert only one option: crash land behind enemy lines.

He wanted to line up a road or a field, anything mildly straight and flat but he could only see shadows in the darkness. His plane was badly damaged and some of the controls were shot. All he could do now was hold on and pray. He didn't want to give up on William who was helplessly depending on him. Rupert envisioned a large open road that he could calmly land his plane on. He dropped in altitude, trying to steady the controls. The left wing was burning, and the tail rudder barely worked at all. His knuckles turned white as he grasped the controls, steadily bringing the plane lower.

I'm so low we must be under the tops of trees, if there are any here, he thought. The night was dark. The flares in the distance created a dim light but it wasn't enough to see much where he was. The trees he could see looked like the shadows of ghosts and ghouls playing in the wilderness. Still lower they went. The plane jerked to the left so Rupert instinctively pulled to the right and he felt the wheels finally touch the ground. The German planes flew past overhead as if taunting them. Rupert tried to keep his damaged plane steady as it bounced along the makeshift runway, half driving on the ground, half flying inches above it. He landed the plane down but hit a bump which tossed them up. When the plane hit the earth again the wheels broke clean off and the burning wing snapped at the base, left hanging on by a thread.

They slid out of control. Rupert threw his hands up to shield his head as the plane hit the trunk of a large tree. He desperately held onto consciousness and climbed out of his seat. He called out to William but was answered by silence. Rupert jumped to the rear seat to pull William from the wreckage. He wrapped his arms around his friend's body and pulled with all his might. He dragged William away from the plane as the fire spread from the wing to the plane's body. He fell to the ground 25 yards away, unable to pull any farther. His hands were covered in blood, and he realized William was bleeding profusely.

"Wake up William!" Rupert pleaded. He ran back to the cockpit hoping to find the first aid kit and some supplies, but the plane was now engulfed in flames. Rupert turned back toward William but as he stepped, his leg gave out and he collapsed.

"When did I get hit?" He desperately blurted out as he crawled back to William. Rupert was exhausted but pushed through. I let all my men die, he thought, I can't let William die too. He got back to William and attempted first aid. He remembered his training. I have to stop the bleeding, he thought.

He opened William's leather jacket exposing a large bullet wound in William's chest. It was too late; William was already gone. Rupert fell backwards, devastated. The last few minutes of combat replayed in his mind and he realized William had not said anything. He died before the crash, Rupert realized in complete despair. The last thing William had done in life was fire at the Red Baron. [FACT 19]

Tears started running down Rupert's face. He was physically and emotionally overwhelmed, hunched over his best friend. Rupert had nothing left to fight for. To his knowledge, he was the sole survivor of his whole squadron.

"I let them all down," he said, "and I have nothing to show for it." Rupert used all his might to heave himself up against a tree. His vision began to fade, he wasn't sure if he was dying or passing out, but in that moment, he didn't care either way.

FACTS

17.   Lieutenant Robinson received the Victoria Cross for successfully destroying the first zeppelin over Britain just two days after the events transpired. He did not receive the medal at a forward base. Instead, he was awarded the nation's highest medal at Windsor Castle in front of a massive crowd including royals and nobility. [Ref 12]

18.   When Leefe got back to the warfront in April of 1917, the average life expectancy of an English pilot was less than 20 hours of flight time. This became known as Bloody April. [Ref 13]

19.   Leefe was moved to the 48th Squadron on the Western front in March of 1917. It was on 5 April during his very first real mission, that he came face to face with none other than the Red Baron. Leefe led six fighters, four of whom were shot down, including himself. [Ref 6, 10]

The Somme Offensive

The narrative was based on the Somme offensive. Leefe was shot down in the same region but it was after this offensive ended. The following is a general timeline of the actual events of World War I.

Timeline of Actual Events [Ref 12]

June 28, 1914 – Assassination of Archduke Franz Ferdinand

July to September 1914 – German offensive, halting just East of Paris

September 1914 – Battle of Marne – Allied Victory pushing Germans away from Paris

January 1915 – Zeppelins bomb English coastal towns for the first time

March 1915 – Lt. Robinson is assigned to the French warfront where he is injured

May 1915 – Zeppelins Bomb London for the first time

May 7, 1915 – The sinking of the Lusitania, increasing America's involvement

February 1916 – Verdun Campaign begins (Allied Effort)

July 1, 1916 – Battle of Somme begins (Part of the Verdun Campaign)

September 3, 1916 – First zeppelin shot down (by Lt. Leefe Robinson)

November 18, 1916 – Battle of Somme Complete (no major achievements)

December 18, 1916 – Verdun Campaign ends (550,000 Allied / 450,000 German casualties)

March 1917 – Lt. Leefe Robinson is reassigned to the French/Belgian warfront

April 5, 1917 – Lt. Robinson is shot down and becomes a POW

April 6, 1917 – US Declares war on Germany

June 1917 – General Pershing leads American Expeditionary force into France

November 11, 1918 – WWI Ends, all prisoners of war are released

Chapter 8

His Castle in the Sky

Rupert could barely open his eyes in the morning light. His head was throbbing in pain. He felt the rough bark of the tree he slept against and the memories came flooding back. He was muddied and bruised, and he was sure he was bleeding from somewhere on his face, but his hands were asleep. I must have slept on them all night, he thought. He didn't try to get up, he just sat there, defeated, and hopeless.

He adjusted his legs but a white-hot pain flared up from just above his knee. He remembered how it gave out when he was trying to walk last night. He reached for it but could not move his arms, he didn't have the strength to try. Rupert quickly relaxed his leg and slid off the tree, resting his head on some grass, looking up at the sky.

My arms must be broken, he thought, actually my whole body is aching in one way or another. The morning sunlight crept slowly upward in the otherwise pale blue sky as Rupert had neither the strength nor the will to do anything. He squinted and saw a few light clouds drifting peacefully overhead. How nice it must be, he thought, able to just float there, totally unaffected by anything we are doing to each other down here.

"Bringen sie den piloten hierher," demanded a rough sounding voice from somewhere in the thicket of woods. I was shot down behind enemy lines, he remembered, if the Germans find me, they will probably kill me. He was too weak to do much, he could barely open his eyes and when he did his vision was blurry. He dragged himself to the other side of the tree, in a desperate attempt for concealment. His mind raced through the events of the night before: the assault on the artillery, his squadron, the Red Baron, William.

Out of nowhere, loud footsteps clomped their way through the mud straight towards Rupert. He instinctively kicked his good leg at the approaching silhouette in the dark uniform, but it was useless. A pair of hands grabbed the front of his jacket and threw him onto his feet, forcing him to walk.

Rupert opened his eyes enough to see a German soldier, dirty from a long fight. He tried to fight the soldier but noticed something rough scratching his wrists and hands. His arms weren't asleep or broken; they were bound. He had already been found, tied up, and made a prisoner. With a fresh shot of adrenaline enhancing his senses, he looked around searching the area for his plane or for William, but he found neither. He had been moved to another location while he was unconscious.

He took another step with his right leg and fell to the ground; the pain overpowered his adrenalin. The soldier took this opportunity to punch Rupert in the stomach before pulling him back to his feet. Rupert didn't want to continue but his feet kept shuffling along with what little strength they could muster.

He was led to a large, armored vehicle with small windows. The soldier threw him in the back through a thick armored doorway. He fell to the floor with a thud that reverberated every corner of the metal box he was now confined to. Rupert felt like a wounded dog caught by a nasty dog catcher. He looked around, frantically trying to figure out where he was or where he was being taken. He scurried up onto the cold bench seat of this otherwise empty mobile prison to peak through one of the windows. From this angle, all he could see were the tops of evergreen trees. He looked frantically back towards the door.

He wanted to see some sort of landmark, but his vision was blurred. With his arms still tied behind him, he fell off the bench back onto the dirty, cold floor. He concentrated on the open door, trying to focus his eyes. A silhouette appeared, blocking the light, coming directly towards him. He squinted and focused his eyes just in time to see the butt of a soldier's rifle as it knocked him clean out.

As he came to, Rupert's head throbbed with a pain so unbearable it was hard to breathe. The rusty hinges on the door screeched as a German soldier swung it open. Blinding light poured in. Two sets of hands grabbed Rupert and pulled him out of the vehicle. He immediately lost his footing and fell to the muddy ground.

He was ashamed of his weakness. I can't even carry myself, he thought. Rupert worked hard to come in first place in everything he did. He had been a winner for as long as he could remember, losing only to his older brother. In school, his excellence was recognized, and he was made head-boy of his house but here on the battlefield, where it really counted, he couldn't even carry his own weight.

The soldiers got him to his feet and marched him toward a long black train. Rupert was tossed into a cattle car as a German soldier barked orders and another slid the huge door closed. The latch engaged with a loud clang, and the inside of the train car was covered in darkness. Rupert took a moment to comprehend what was going on around him. There were other people with him. If their language was any indication, four of the men seemed to be French and two were British. All seven men were now prisoners of war, having been thrown on a train like livestock being sent to slaughter.

The train hissed and cranked to life. The dirty train car chugged its way to their unknown destination. It was stifling and became hot in the spring sun. After his eyes adjusted to the darkness, Rupert saw two other British soldiers were also injured but their wounds had been bandaged.

"Look at your leg, Sir!" One of them said to Rupert before turning to the other soldier, saying, "Davies, have you got any bandages left?"

"The Germans took everything I had," Davies replied.

"Well, we need to do something, he's an officer for God's sake," he said referring to the two small oak leaf insignias pinned on Rupert's shoulders.

Davies took off his overcoat, then ripped a strip of material from his own shirt to create a bandage. He wrapped it around Rupert's leg while the wounded soldier supervised.

"Make sure it's tight!" He instructed.

"It is." Davies said impatiently.

"And put some pressure on it!"

"I am!"

"Don't tie the knot there, it will slip." The wounded soldier instructed.

"Why don't you crawl over here and do it then?"

"No, you're doing great... but you do need to move it down a little."

"I know what I'm doing! I bandaged you up, didn't I?" Davies said, now thoroughly annoyed.

"You mean this?" The wounded soldier said pointing to a loosely wrapped bandage around his arm. "It's fallen off twice already!"

"Only because you keep messing wif it!"

"No matter that now, you just help the Leftenant."

This small act may very well have saved Rupert's life. He wanted to thank Davies for bandaging him up but between the pain and the heat, he felt sick to his stomach. If he opened his mouth the only thing to come out would be vomit. The train rocked back and forth as it jogged down the track. He managed to contain his queasy stomach but started to feel light-headed before his vision, and his pain, gradually faded. Just before darkness overtook him, Rupert heard someone say, "You don't look so good Sir."

Sometime later, the train's large door slid open, clanging violently. The car shook angrily, and Rupert awoke to German soldiers standing in the doorway. Some had rifles and others just shouted at the prisoners, forcing them off the train and through a fenced gate. Rupert could barely walk; he fell to his hands and knees just inside the fence. Totally unaware of his surroundings, he feared if he stopped the Germans would beat him or just leave him for dead, so he crawled as far as he could. With his limited vision, he saw a dilapidated building and a few dozen people walking towards him.

I can't go any further, he thought is despair. He flopped to the ground and rolled onto his back. The people surrounded him. Rupert couldn't discern who they were, but they towered over him as faceless silhouettes against the overcast sky. I'm being fed to a pack of wolves, he thought. With his last ounce of strength, he prayed, dear God, please let me to make it back home. With the thought of his seaside village in his mind's eye, he lost consciousness once again and his desperate prayer was released into the primordial energy of creation as if being received by God.

Fortunately for Rupert, they weren't wolves, or Germans. He had been delivered to a prisoner-of-war camp and the silhouettes were prisoners who came out to see if they recognized any new arrivals. After seeing Rupert, one of them bravely harassed a guard for some medical supplies but his efforts were in vain.

Days turned into a week and Rupert remained mostly unconscious, waking only to drink a little water and eat what food the other prisoners could feed him. They had very little food and water, but they shared the best they could. Rupert became a symbol of hope for the prisoners who were all rooting for him like the underdog in a fight against evil. Overall, he didn't seem to be getting better but he wasn't getting worse either. That changed after another few days as the dirt and dust infected his leg. This time there was no blue glow to save him.

The guards conducted routine inspections, walking through the living quarters, mostly just to harass the prisoners. However, this time Rupert's infection was so bad the smell made them gag. They called the prison's doctor. He immediately ordered Rupert to the infirmary where he unceremoniously amputated the festering leg. There was no consultation with Rupert or any attempt to control the pain of the amputation. Luckily for Rupert, though, he didn't feel a thing because he had fallen into a very deep, coma-like sleep.

The next morning, a cold rainstorm blew in, bringing a steady, dreary rain. The guards trudged through the fresh mud to conduct some sort of inspection. They got to Rupert and annotated something in a small notebook.

A few hours later a train arrived bringing more prisoners. This time, instead of chugging off immediately, the guards loaded a few severely wounded prisoners into one of the cars. They got Rupert to his one remaining foot, but he didn't understand what was happening. Instinctively, he fought the guards from his drunken-like stupor. The guards simply dropped him into the train, letting the hard floor dish out the punishment.

Rupert wouldn't remember fighting so hard that the guards dropped him. He wouldn't remember much of the train ride either.

A man, maybe around 40, with short hair and a thin mustache, walked up to Rupert as he was being set down in bed. He marked something on his clipboard.

"Where am I?" Rupert blurted out while trying to open his eyes for the first time since the guards dropped him in the train. He looked around and saw a large open room filled with metal beds, all covered in white linens, arranged with precision like soldiers standing in formation.

There were sick or injured patients lying in most of them. "A hospital," he answered his own question. He glanced down and saw he was still in his filthy uniform.

After living in shoddy plywood buildings for the last couple years, Rupert felt like he was in a proper hospital but, in reality, he was in an Army field hospital, converted from something else after the war started. His bed was one of four that were tucked into a corner of the open room.

"Why, my boy, you are in Switzerland!" the Doctor with the thin mustache said through his thick Germanic accent.

Switzerland, Rupert thought, weren't they neutral? Had they joined the war on Germany's side? [FACT 20] Rupert's missing leg felt like it was on fire, but his mind was overcome with the revelations of this apparent German-Swiss alliance. He thought about how they could, or maybe already had, attacked France and created an entirely new front in the war. There's no way the Allies can fend off attackers from the South as well as the East, he worried.

The world around him began fading away. The blinding sunlight reflected off the white linens, becoming blurry resembling the puffy clouds he used to fly through. This was different. He was not flying. It felt as if he was approaching heaven's gate. He expected an angelic chorus to sing out any moment. Before he could ponder what was happening, he began to descend from the clouds. He saw visions of his peaceful village back on the English coast. He was desperate to be back there. He saw green pastures ripe with crops under a brilliant blue sky filled with the giant puffy clouds he had just been floating in. Now, firmly on the ground the clouds resembled dreamy pillows floating peacefully through the heavens. The sounds of people groaning in pain calling out for a nurse brought his focus back to the hospital ward. In that painful moment, Rupert did not understand why he left his homeland for war in the first place.

Rupert found himself in a strange place, like a dream, only it felt more real than anything he had ever experienced. At first, the pain in his body lingered until it melted away and he wasn't bothered by it at all. He started to feel things he never imagined feeling. He was more aware than he had ever been. He was standing, or perhaps floating, outside some old building. It was a plain one-story structure, nestled in some foothills below giant snowcapped mountains.

He could tell it had been a stagnant place but was now bustling with fresh energy. There was a green spade-shaped leaf on a vine growing near one of the windows. He focused his attention on it and felt the wind as if he was the leaf, like an out-of-body experience but instead of seeing himself as Rupert, he was this leaf growing on the vine. He watched the leaf shake in the wind and felt his whole leafy body flutter in the breeze.

He blinked and the building was gone, he was in a new place entirely, standing in a small, manicured clearing in the middle of a forest. There was plenty of light to see in any direction but the whole place felt dim as if it was illuminated by nothing but the light of a full moon. The clearing was carpeted with soft grass that was green and perfect. The edge of the grassy clearing gave way to the surrounding woods which encircled the whole scene. Rupert peered into the forest. The woods were not too thick, so he should have been able to see further but instead they faded into a dark nothingness.

It was the most peculiar place, as if it was an island in its own reality, somewhere outside of ours. Rupert turned to his left and realized the clearing curved around like a large L shape. He noticed a few tulips blooming in a small garden bed running along one edge of the perfect lawn. He followed the garden to the bend of the L-shaped clearing where large shrubs and invasive vines grew unchecked. Rupert stopped, something felt odd about that spot. He pushed the vines away revealing a grand, stone staircase that

felt old, possibly even ancient. Rupert sensed the stairs themselves knew all
the hidden secrets of that place. The steps led up and out of this dimly lit
clearing in the woods.

To the left of the stairs was a small tree, not much taller than he was.
Just as Rupert focused on the tree the strangest thing happened. His vision
magnified on the exact spot where he was looking. He could see the most
intricate details of the tree; it was like he was looking at its individual cells. He
realized he could understand the tree as if he was the tree. The place felt
peaceful, but Rupert understood he could not stay there for long. He walked
towards the stairs and noticed a drip echoing from the direction he had just
come from.

He turned and found a stone pedestal holding a bowl of water like a
birdbath. He peered into the water and before he saw anything he was
transported to a cave-like place that felt as if it was below, or perhaps *inside*
the clearing in the woods. Rupert was now floating on the surface of a dark,
calm ocean that seemed to go on forever in every direction. This place was
dimmer than the woods, but he could see further, though there was nothing
to see. The soft acoustics of the clearing were replaced by a harsh echo as if
he was in a rocky cave. There was some sort of ceiling but there were no
walls as far as he could see.

Rupert understood he was inside of himself, in a vast ocean of his
own soul, and yet, he was powerless to do anything other than float there. He
turned and looked in another direction along the surface of the water.

His plane was sitting on top of the water as if it were resting on dry
land. It was badly damaged and still on fire, just as it was the last time he saw
it before it was totally engulfed by flames. One wing was almost broken off
with the tip resting on the ground while the base of the wing was hanging
onto the body of the plane by a thread. The fire spread and cracked as Rupert
closed his eyes and shuddered. His plane had been an extension of himself
and in that moment, he knew he would never fly again. He was unable to
look away, so he tightly closed his eyes.

Something changed so he opened his eyes, and realized he was back
in the peaceful clearing in the woods. He was standing next to the bowl of
water shimmering in the soft light, though he did not dare look again.
Instead, he looked to the ancient staircase and saw a heavenly light radiating

from somewhere above it, spilling its warmth down the stairs. He tried to look past the overgrown shrubs and vines but he couldn't see where the stairs led. He stood peacefully, being showered by the glow for a few moments. It felt peaceful, like home. It felt right so he decided to climb to the top, just to see what was there. He looked up the stairs and saw a landing every few steps where the whole staircase would change direction, creating a spiraling pattern. In total there were maybe two dozen steps and four or five landings.

He started climbing. After an extraordinarily long journey to the second landing, he paused and looked back down at the humble clearing in the woods. He noticed things he hadn't seen when he was down there. He could see a bench he recognized from his parent's coffee plantation in India and beside it was a football he used to play with when he was a boy.

Something about the stairs didn't seem right. He was much higher than he expected to be. He counted the stairs he had already climbed, there had to be three dozen, at least. How strange, he thought, there were way less stairs looking up from below.

He looked up and two landings away, he could see where the stairs were going. They led directly to an arched doorway in the sky; there was no door though, it was just an open archway. He could see through the opening into a room that looked like a chamber in a castle. It was a big open room, and the light was emanating from something inside.

How odd, Rupert thought, that room has no supporting structure. The stairs he was climbing were made from old stones, and so was the inside of the room but from the outside, the room didn't seem to exist. The stairs and archway were surrounded by nothing more than the light blue evening sky.

He continued climbing the stairs. Farther and farther up he went, but the bright light kept pace with him, never getting any closer. He stopped at the next landing to look down the stairs and saw the dozens and dozens of steps he already climbed. He looked up the stairs, which turned sharply and became steep. The archway was still two landings away. It isn't getting any closer, he thought in frustration, and the climb is getting harder.

He felt the warmth of the light. It felt like love itself. Rupert let out a long deep sigh and finally admitted to himself a truth he was trying to ignore. I am dead, he acknowledged. He looked at the stairs he was on and knew he was walking towards heaven. It was God's light shining down upon him. It felt so warm and loving, he wanted to bask in it forever.

I'm not just going up to look, he thought, I'm going to go through that door. "I'll do whatever it takes to get there." Rupert stated as he began climbing towards the archway at the top. "I'm ready to go into the light." With this decision he started to feel weightless; he effortlessly floated, no longer burdened by the climb, he was delivered to the final landing in front of the archway. He could see the sky surrounding the portal and for the first time he could also see the structure that supported this castle in the sky. The ancient stones only revealed themselves slightly, allowing a momentary glimpse, like the rainbow shimmer of an oil slick. The stone masonry of the large castle walls acted like an invisible force field floating in midair. Somehow it was a castle, complete with turreted towers and medieval tapestries while at the same time it was hiding in plain sight, disguised as the sky itself.

Rupert slowly walked through the archway, leaving the stairs in the sky behind him. He entered a large cold room. There was a massive, ornately decorated door which was closed but the cracks around it allowed bright strips of glorious light to escape. The light seemed to flow outward in every direction. For a moment he enjoyed the comfort of the light. This is the door to heaven, Rupert thought to himself, when I go through this door, there will be no turning back.

FACTS

20.   Switzerland remained neutral throughout both World Wars. They did, however, engage in humanitarian missions by accepting prisoners of war who were 'found' inside Switzerland's boarders. Switzerland created four POW camps which acted as hospitals where they housed and nursed troops before sending them back to their countries of origin.

The camps were split by nationality to avoid any conflicts amongst the prisoners. Prisoners were only sent back to their country if they had sustained some injury which would prevent them from rejoining the war. [Ref 14]

Chapter 9

The Ghost in the Machine

"Absence of evidence is not evidence of absence."

-Carl Sagan,
American Astronomer

Rupert stood, basking in the light that was escaping through the cracks around the giant door separating him from heaven. The light blanketed him in the most comforting feeling imaginable. But wait, Rupert thought, if heaven is through the door, where am I now? I'm certainly not among the living. This is no hell or purgatory I've ever imagined; it's like heaven's waiting room. I'd like to look around, he thought, after all going through that door will be a one-way trip.

Rupert slowly turned in a complete circle, starting to his left, towards the stairs and the clearing in the woods below. He kept turning, now facing away from the door of light. He saw a large opening in the stone wall that revealed a wide spiral staircase inside a turret. This surely must lead to some tower, he thought. He continued turning and saw a dull and tarnished suite of armor that looked as though it would not be very effective in battle. He finished turning, now with the heavenly door to his left, he saw a large dark corridor.

It was a peculiar hallway with ceilings so tall he couldn't see the top. He was drawn by something at the other end. He crept down the black hallway and the walls themselves seemed to dissolve into nothingness. As he walked through the darkness he began to wonder if there were walls at all, so he reached out and ran his hand along the side of the corridor. What he touched, moved, like a giant curtain. Three thick ropes dangled out of the darkness from somewhere above. He felt that if he pulled them in a certain order then the curtain would open, revealing some grand play.

He brushed passed the ropes and the dark hallway opened to a chamber where a lit torch marked the entrance to yet another room. This room had no door, instead, the open archway led to a large, mostly empty, space where a cozy living room, complete with a fireplace, was tucked into the far corner. Rupert walked up behind the two large armchairs facing the crackling fire and noticed the unadorned mantle against a brick wall. There was a rug on the floor that created a very pleasant room tucked in the corner of the otherwise open space.

"Come, sit with me, won't you?" an old man's voice startled Rupert. Someone was sitting in one of the chairs.

Rupert was alarmed that someone else was with him in this odd place; but the stranger felt familiar. Rupert moved towards the fire and turned around. There was an old man sitting in one of the chairs. He stared in amazement, trying to figure out who this person was. The old man had been reading a book but closed it as Rupert walked in. He set it on the side table and Rupert saw the book's cover and read the title, *The Life of Rupert Robinson*.

"What is that?" Rupert asked with a mixture of concern and intrigue.

"Oh this? Just reminiscing on the life I lived." The old man said.

"Are you… me?" Rupert said weakly.

"Yes, and no." the man replied mystically.

"But I'm dead now," said Rupert, "aren't I? You are old so you must be a version of me that I never got to become."

"Actually, I am not you at all, but you are me." The old man replied behind misty eyes.

Rupert didn't understand and did not hide his frustration, but he let the old man continue. "You see, our ancestors live on in each of us. My name is Rupert Robinson. Your father named you after me."

The Ghost in the Machine

For the first time ever, Rupert met his grandfather who had died when Rupert was just a baby. Rupert was overwhelmed with emotion and confused, but in this odd place, seeing his dead grandfather seemed to fit.

"You chose me to be the one who helps you to the other side." His grandfather said.

"What do you mean, I chose you?"

"It is time for me to show you the way home." His grandfather said, slowly raising a hand, pointing in the direction Rupert had just come from.

Rupert turned expecting to see the dark hallway but instead, he saw the door of light; they were only a few feet from it. He could once again feel the warmth from the light. His grandfather picked up the knocker that hung from the center of the grand door and knocked three times. Each knock reverberated a sound that Rupert couldn't hear and yet he felt throughout his entire being. The door was like a clock or a time machine that had recorded every part of Rupert's life; and knocking on the door projected the recorded episodes of his life into reality.

The first clang transported Rupert to his own birth and early years. He started reliving his life, but he experienced each interaction from other people's perspectives. He could feel his mother's happiness when he took his first steps. He could feel the shame and anger he caused a young boy he played a trick on at the coffee plantation back in India.

With the second clang, Rupert was transported to 1913 when he was formally being appointed as Head Boy of Eaglesfield House at St. Bees School. He reexperienced the pride he felt as he stepped into the same role his brother held two years before him.

Rupert also experienced the emotions people felt as they watched the ceremony from the audience. Some looked on in admiration that such a noble family graced their school. Others were envious as they felt they were more deserving of that honor. For just a moment Rupert was shown a dusty trench in Babylon where he saw his brother reading a letter Rupert had written him. Rupert felt his brother swell with pride as he read that Rupert was the new Head Boy. Rupert had never gotten confirmation that his brother received that letter. His brother died in battle soon after.

The third clang of the knocker against the great heavenly door rang out and Rupert saw himself at his commissioning ceremony when he became an officer. They pinned the emblem on his pressed uniform, and he felt his father's pride. He relived every minute of his life in the military. The whole way, right up to William's death. He got to see everything from a new perspective, somehow outside of his own experiences. He watched himself evading the Red Baron. Before they crashed, he saw William get shot by the German fighter, confirming his suspicion. William died before the crash and Rupert could feel his final thought; William was praying for Rupert to make it home alive.

The sound of the final knock dissipated, and Rupert's life review was complete. The door swung open and a blinding light filled the entire space. Rupert felt more wonderful than he could have ever imagined. It really is God's light, he thought. There was no other way to explain what it could be. Still, some part of Rupert wished he didn't have to die.

He turned to his grandfather and asked, "do I have to go?" Rupert had to fight to hold on to the desire to return to life because the light felt so good. He knew all his pain and problems would simply melt away if he just went through that door.

"It is always your choice." His grandfather said. "There is no right or wrong; only a choice and the consequences that it creates; for better or worse."

Rupert stood still in that ethereal realm, outside of time and space. He watched his grandfather step into the light and disappear. His grandfather spoke again from beyond, though it sounded exactly like it did when he was standing right next to him.

"All you have ever had is a choice and the consequences it created; for better or worse." His words echoed through every inch of the castle's halls.

Rupert wanted to go back to his earthly self, but the light drew him in like a magnet sliding him right up to the threshold but it wouldn't pull him through. He had to choose. He stood, swaying on the precipice. He knew stepping into the light was a point of no return. He managed to look away, back down the staircase to the small clearing in the woods where he had first appeared in this strange place. He thought about the home he left back in England. He wanted to go back to it.

The Ghost in the Machine

He realized he wasn't just looking at the stairs. His perception seemed to be physically floating towards them. He looked back to the heavenly door but caught a glimpse of the dark hallway that had led to his grandfather. The ropes that were hanging from the ceiling were now moving; two fell to the ground and one raised until it disappeared. The curtain covering the wall was opening and a movie began to play. He saw a doctor holding bag of red liquid with a tube attached to it.

He wanted to go watch but he no longer controlled his movements, his consciousness was retreating through the archway, out of the castle and past the first landing. Instead of seeing the small, wooded clearing, he descended into a small, cluttered office and just like that, the castle with the door to heaven was gone.

The office had a desk where Rupert saw a large book, bound with a red leather cover. The book was open to a page with a magnificently illustrated letter D taking up a good bit of the page. The D was surrounded with odd symbols but most odd of all was that right in the middle of it, was a medieval castle. He was overcome with the sense that this was the castle he just left. For a moment he lingered on the page and studied each symbol.

He reached out to touch the book but instead of feeling a piece of paper, he felt nothing. He started hearing the sounds of people in pain and nursing staff rushing to help the injured. He saw white sheets drenched in blood. Rupert focused his attention on this and the office disappeared, replaced by a hospital room that Rupert was floating in.

"Die bluttransfusion verabreichen!" He heard a voice say. He didn't know the language but somehow, he understood exactly what was being said.

He followed a nurse into an operating room where he seemed to bounce off the ceiling as if he was a balloon filled with helium. He saw a lifeless man, mostly covered in a sheet, lying on a bed surrounded by nurses and doctors. They were holding up a bag of thick, red fluid and stuck the man with a needle, pumping it into his body.

The doctors and nurses parted, allowing Rupert to finally see the man on the bed. It was his own lifeless body. He was somehow a ghost in the room watching his own treatment. Two doctors with charts furiously scribbled notes and shouted directions to the nurses. Nursing staff frantically wiped up spilled blood while manually taking his blood pressure and pulse.

The doctor with the thin mustache that Rupert had briefly spoken to was calmly standing in a corner, watching it all happen.

Everyone was fussing so much over him; it seemed almost silly now. After basking in the glory of God's light, nothing seemed to be too serious, not even his own death. A nurse walked out carrying bloodied sheets, then another nurse with clean sheets walked in.

All of Rupert's senses focused on her. She was beautiful but Rupert could see something more. There was a connection he recognized from sometime long ago, as if they had crossed paths as children. He noticed something different about her, she had changed since they last met, though he couldn't tell how.

From his ghostly existence her face bore the resemblance of someone who had experienced true pain and sorrow and yet she seemed like an angel who was capable of healing anyone. For the first time since he entered this ethereal dream, Rupert could feel his body. At first it was just his chest, where his heart beat with a warm curious love for the nurse.

Rupert was drifting closer to his body as the doctor with the thin mustache walked up to see Rupert over an attending nurse. Rupert looked at himself and recognized an agonizing pain and then it happened. All at once he felt pain all over his body. Every bit of him ached and he groaned in agony. Everything went black and Rupert could feel his right leg; it felt like it was being electrocuted with no reprieve. His head throbbed before the whole scene dissolved into darkness. [FACT 21]

Rupert had been at the hospital in Switzerland for three days now but had only been conscious for a total of a few hours. His missing leg still throbbed with pain, which was easy enough to endure. Worse was the understanding that he was missing a leg and was forever changed.

Over the next several days he began to show signs of recovery. He was able to spend more time awake and eat small bits of bread. He didn't care much for the fermented cabbage they fed him but he tried eating it anyway. Within a few more days he even managed to stand on his good leg for a few minutes. It was in this time that the doctor with the thin mustache appeared at his bed and barraged him with questions.

The Swiss doctor spoke in a heavy Germanic accent, "My boy, I believe at some point, you may have actually died here in this hospital. It was because of that dire condition that I authorized an experimental treatment."

"You gave me someone else's blood." Rupert said in a matter-of-fact way. He didn't bother looking at the doctor, he remained in his bed with his eyes closed, focusing more on enduring his pain rather than whatever questions the doctor was asking.

"This is correct!" The doctor was surprised. "How did you know this?" Rupert's ghostly journey flashed through his mind, but he stayed silent for fear of sounding insane.

"You see, we received human blood from a donor and we injected it into your veins as if it were your own. What is very curious is that you were able to receive the blood where others have died, or worse."

"What's worse than dying?" Rupert mumbled without thinking.

"For some, the transfusion seems to cause agonizing pain. Usually, the patients were in pain before the transfusion but after they received the blood they convulsed until they died. We do not know why you were able to receive the treatment while others were not; you are very lucky." [FACT 22]

Rupert said nothing. He was still in immense pain, missing a leg, and didn't exactly feel lucky.

The doctor asked a litany of questions and took notes on his clip board. He was unlike any doctor Rupert had ever spoken to. He asked very specific questions about Rupert's physiology like, if his lungs and heart felt normal, but he also asked peculiar questions about his thoughts and feelings, and it seemed like he took more notes for this type of question than the medical ones. Sometimes he would bring up dreams but Rupert would never entertain those questions. For Rupert, the reality of that experience was too strange to talk about, even if it was just a dream.

Time went on and Rupert's health improved by leaps and bounds. The memory of that dream or whatever it had been, often dominated Rupert's thoughts. If anything, he felt a renewed sense of connection to God, but what is a dream anyway? He pondered. Should I ask the doctor about it and who was that nurse in the vision? It had been nearly two weeks since Rupert woke up and he had never actually seen her in person. Had she been an angel disguised as a nurse?

Sometimes, he could relax and live, what felt like, a normal life but normal now seemed as strange as being in any foreign land. After all, he was trying to find normalcy in a field hospital while learning how to live with one leg.

The hospital itself was in an old building whose original purpose was lost to time. The rectangular-shaped structure had a wall down the center, dividing it into two big rooms. Hospital beds lined each room creating pathways around the floor. One corner of the building pushed further out than the rest, not keeping form with the rectangle. This created a cubby that was almost like a room unto itself. This is where Rupert's bed was, along with three other patients.

The old building was at the top of a rolling hill, surrounded by walking paths. The hill seemed like an anthill compared to the Swiss Alps that jetted up out of the Earth and towered into the sky above. The peaks were covered in pristine white snow.

The walking path meandered around the rolling hills where there were three other buildings that also belonged to the hospital. There were living quarters for the staff, a warehouse for supplies, and a small building hidden by some trees that the Doctor with the thin mustache had taken up as his office and residence. The path turned into a road that led down the hill to a nearby village which sat in the flat valley between mountains.

During his normal waking hours, Rupert found comfort in enjoying the natural beauty of the place. He focused his mind on things that would allow him to forget his troubles.

He found an English version of a novel written in the 1860's by Wilkie Collins, called *The Moonstone*. He also found the game of chess to be a thrilling part of his day. Someone once offered to play Crown and Anchor, but it brought back too many painful memories for Rupert.

All things considered, Rupert knew he was lucky to be off the battlefield, but he couldn't help feeling guilty. He wished he could somehow do his part to help his fellow brothers in arms. He craved hearing news of the war's progression and began writing to all his chums back at Sutton's Farm. Without a leg, Rupert knew he would never return to battle and yet his own war was far from over.

FACTS

21.   The current scientific understanding of consciousness is limited to our waking experiences. The argument is; there is no evidence to suggest otherwise but as Carl Sagan and Martin Rees explained, an absence of evidence does not imply evidence of absence. The thing is, there is evidence (a consistently observed phenomenon). Some form of consciousness, not reliant on the physical brain has been observed in nature.

NDEs, or Near Death Experiences, are a well-documented phenomenon that mainstream science is beginning to take seriously. It is the consistency that is so intriguing. People from every culture and religion across the globe experience this form of consciousness-outside-the-physical-body. There are even historical records indicating that people have been having these experiences throughout all recorded history. When people try to describe their experiences, they use the verbiage of their culture and religion but when we look at each description objectively, we see a clear pattern. NDEs are an experience that transcend religion and culture.

Rupert experienced an NDE that contained typical elements reported by most NDE survivors; he met a deceased family member, had a life review, basked in a glorious white light that could only be described as love, saw his own dead body and the doctors fussing over it, and experienced things away from his physical body that helped to later corroborate his experience. The real-life Leefe did not report an NDE.

Devoted Christians, Pagans, Muslims, Jews, and atheists have all reported NDEs that follow this pattern. Most come back to life with a personal understanding that their religion is inconsequential.

Strictly speaking, this does not scientifically indicate an afterlife. Since the physical body was resuscitated, we cannot rule out the physical body had some involvement with the NDE. What we can scientifically state is our individual consciousness can exist outside the physical body. This indicates our consciousness is not within our brain which would explain why science has never found physical evidence of consciousness.   [Ref 15, 16, 17, 18, 88]

Fact 109 provides more information about NDEs.

22.   The first successful blood transfusion in humans took place in 1667, however the practice first used animals' blood. While some of these

experimental transfusions were successful, some resulted in agonizing consequences to include extreme pain and black urination. The practice remained misunderstood until blood types were discovered in 1907. World War One acted as a catalyst for the technological advances of blood transfusions. Blood stabilizing agents were invented, giving blood a shelf life. Before this, blood had to go directly from donor to receiver. [Ref 19]

Chapter 10

Kindred Spirits

"Nurse, could I have some water?" Said a patient in the fresh sunlight of an early morning.

"It's her!" Another whispered loudly.

"Nurse, could you help me get to the washroom?"

"A blanket please, nurse!"

Rupert awoke one morning to the sounds of patients calling out for a nurse. I've never heard such a commotion in this hospital, and what silly inconveniences, he thought as he rustled in his bed trying to sleep.

Some were exaggerating their pain while others came up with any excuse to call this nurse's attention. The sun was shining directly on Rupert's face, so he gave up on extra sleep and just laid there trying to forget his pain. He really was thankful to feel the sunshine of another day.

He slowly opened his eyes and peered out from his corner. That's when he saw her. His heart skipped a beat as if two shots of espresso hit him all at once. His eyes focused on her and everything else around her faded out of existence.

It was the nurse from his vision. She wasn't an angel after all, he thought. She was even more beautiful than he remembered. She had blue eyes and long dark hair tied tightly behind her head.

She spoke with an Eastern European accent, "I remember you," she said as she walked into the cubby of four beds tucked in the corner where Rupert was lying. "You got a blood transfusion right before I went on leave. I am happy you have made it. The doctors were not sure if you would live."

"Oh, yes, I… I am here, and so are you…" Rupert said awkwardly before she walked away to another patient who sounded like he might die of dehydration if she didn't bring him some water right away.

Her mere presence was wonderful. She had the caring touch only nurses have. Between her physical beauty and her gentle, but strong, nature she could have easily been mistaken for an angel. She made each patient feel cared for like no doctor could, but to Rupert, there was something more about her, something that seemed familiar. In a strange way, she felt like home even though she was so novel she wasn't like anyone Rupert was accustomed to. After that brief encounter, anytime she walked onto the floor, Rupert tried to come up with some reason to get her attention just like everyone else did.

"Excuse me nurse, do you have the time?" He asked her one afternoon.

"I am sorry, I cannot reach my watch," she replied hastily, carrying some fresh sheets.

"Oh, that is a shame. You have a watch but no time," he said. "You see, I haven't got a watch and yet I have all the time in the world."

The nurse paused, slightly confused having to translate the play on words then she said, "Clever, what is your name?"

"I'm Leftenant Robinson of the 39th Royal Air Wing, but you can call me Rupert. What should I call you nurse?"

"I am Nurse Sklyar, but since we are being friendly you may call me Sabina."

Sabina was Russian-born and a few years older than Rupert. He found her to be as beautiful as they come. She was slightly shorter than average, with a small button nose and big round eyes that pierced your soul if she looked directly at you. She spoke Russian, German and just enough English to tantalize Rupert's imagination.

It was now January of 1918 and they were in the midst of a cold Swiss winter. Aided by Sabina, Rupert began a rigid physical therapy regimen to adapt to life on one leg. Rupert's humor was almost always lost in translation but somehow he managed to grow close to Sabina. She often accompanied that doctor with the thin mustache whenever he made rounds asking his silly questions but he didn't mind as long as he got to spend extra time with her. Rupert was never quite sure but whenever he was trying to steal her attention, it seemed like she fulfilled his meaningless requests before the other patients. Perhaps she was growing a liking for him as well.

Rupert managed to convince her that he needed to go outside for fresh air, for his health, of course. Once a week, she would bundle him up and push him in a wheelchair along the small path that meandered through the rolling hills around the hospital. Rupert loved the time he got to spend with Sabina without any distractions from other needy patients. The natural beauty of Switzerland made it that much better. The rugged mountains shot up from the ground and rose high into the sky above. They went so high the tips were covered in snow whether it was winter or summer. Switzerland was beautiful, there was no doubt about it, but Rupert could do without the cold.

"This cold is nothing compared to my homeland." Sabina said, as she pushed him down the stone path.

"Remind me to only visit you during summer!" Rupert retorted. "Or perhaps you should visit me in summer. Where I'm from, winters can be cold but summers are absolutely perfect."

"Perhaps one day I will," she said.

Life for Rupert had not turned out as he expected but he tried to find the good in it. Sabina became a beacon of hope, a reason to keep going. Plus, the doctor who always came around asking questions did seem to truly care. He encouraged Rupert to go to college when he got back to England so he could live a good life. Rupert wasn't sure about college, but he often thought about how grateful he was to have been transferred from the POW camp to this hospital.

Rupert was back in the cockpit of his plane. William was behind him yelling, "Get to the German Artillery!"

Rupert gritted his teeth, grasped the controls of the airplane, and sped off in whatever direction William called out. The next thing he knew, there were at least a dozen German fighters surrounding him. They weren't shooting at him, instead they were just shouting and shaking their fists. Rupert began to sweat profusely as they got closer and closer. Still, no one was shooting, but that didn't stop the feeling of impending doom from overtaking Rupert. He knew he was in grave danger. He was shaking with fear but held the controls, taking evasive actions.

The artillery came into view and Rupert yelled to William, "We're almost there, ready your weapons!"

William did not reply. Rupert frantically looked at all the German fighters around them. Every last one was staring directly at him. It was eerie, even their maneuvers were in unison as if something was controlling all of them at once. How are they doing that, Rupert thought.

"Shoot those fighters!" He barked to William.

William didn't reply and the sound of machine gun fire never came. Rupert turned around in his cockpit to see what was wrong, but William was gone replaced by a little girl, maybe 11 or 12 years old, sitting where William had just been.

"William is ok, Rupert," the little girl said calmly.

Rupert was not comforted by this at all as he faced forward to fly the plane. The little girl seemed familiar but why was she saying these things? He glanced back to look at her one more time. She was now standing on the seat totally unaffected by the aerial movements or the wind. She slowly reached out her arm, her hand was closed, holding something that she was trying to give Rupert. He couldn't look away, her arm fully extended, and she opened her hand revealing a scarab beetle which had a green and golden iridescent sheen to it.

Rupert's knuckles turned white as he gripped the controls, he looked forward and saw a giant black raven sitting atop one of the artillery pieces. Some decision was made, though it was not Rupert who had done the deciding. Rupert had to crash his plane into the gun with the raven on it. He was flying the plane and yet he was not in control. He screamed at the top of his lungs as he delivered his plane to its target in slow motion.

He felt the little girl touch his shoulder and gently shake him. "Open your eyes! Open your eyes!" She pleaded.

"My eyes are open!" Rupert yelled as he stared at the target he was rapidly approaching.

He felt cold water hit his face and for a split second he was transported back to that underground ocean from his near-death-experience. He bolted upright and looked around. He was sitting in his hospital bed covered in sweat and cold water. Sabina was by his side holding an empty cup.

"Now your eyes are open." She said coolly in the dark. It was the middle of the night and Rupert's screams had roused half the patients on the floor. "You will have to talk to the doctor about this in the morning." She informed Rupert.

"The doctor who asks those ridiculous questions?" Rupert snapped.

"Do you know why he asks all those questions?" Sabina challenged.

"I haven't the slightest idea." He replied matter-of-factly, forcing a gentle smile.

"He is a psychiatrist. He is a medical doctor, but his specialty is studying the mind." She explained.

"You mean he has been studying me like some kind of lab rat?" Rupert's patience immediately dissolved, and he felt like a prisoner for the first time since he awoke in that hospital.

"He is trying to help you." Sabina explained but, for the moment, it fell on deaf ears.

Rupert tried to think back to individual questions the doctor had asked him. I never stopped to think why he was asking those things. Rupert didn't know what to think, am I a lab rat or is this a good thing? What does the doctor know about me that I didn't realize I was telling him? He asked himself as the thought of that ethereal vision came into the back of his mind like a familiar stranger lingering in the shadows. [FACT 23]

Sabina went back to bed and he put the doctor out of his mind. Rupert's only thought was back to what that little girl said in the cockpit. William isn't ok, he thought. He's dead.

The next morning, Rupert was sitting in bed thinking back to how he ended up in Switzerland and the wild journey that got him there. There were long periods of time where he had no recollection. There was at least one train ride, he thought, a soldier bandaged my leg with his own shirt. Davis was his name, Rupert tried to remember, or was it Davies? I think he saved my life. I feel I owe him a life debt. I hope I can repay him one day, or at the very least thank him.

Just then, the doctor with the thin mustache came quickly around the corner to interview Rupert yet again but this time was different.

"I don't think I ever properly introduced myself." He said through his accent. "I am Dr. Carl Jung. I am in charge of the entire hospital but as a doctor I specialize in psychology, I study the mind." [FACT 24]

"Yes, I had heard that about you." Rupert said a bit defensively.

"Let me give you an idea of what I do." Dr. Jung started, "Are you a Christian?"

"Of course." Rupert replied sounding a bit more superior than he intended.

"Then you are familiar with the ten commandments. Why did God make those rules? Why those ten?" The doctor questioned.

"Because God doesn't want us to do those things." Said Rupert plainly.

"And is that all there is to it?" Dr. Jung asked pointedly. "Do you like breaking the commandments?" The doctor asked without giving time for a response. "Which is your favorite to break?"

"What?" Rupert said rather appalled at the blasphemous thought. "I don't want to break any commandments."

"Nurse?" Another patient called out to Sabina who looked over and saw a patient with bandages on both arms reaching for a pillow that had fallen to the ground. Unfortunately for Sabina, after she picked up the pillow, someone else called for her followed by another and she never came back for the conversation between Rupert and Dr. Jung.

Rupert felt quite vulnerable as he was aware that this discussion was happening on the hospital floor where about a half dozen other patients could hear.

"Why don't you want to break any commandments?" The doctor's tone softened.

"Because God told us not to."

"Right! But what if God never gave us those rules? Then which of those things would you want to do?"

"I don't like this." Rupert said, feeling rather squeamish.

"It is okay to feel uncomfortable." The doctor stated before quickly moving on. "You see, even though God commanded all of humanity to not do those things, people still do them. People who know the rules and know they should follow them."

The doctor was attempting to explore the depths of humanity in all its forms. He spoke as if he knew some fundamental aspect of each of us that Rupert had somehow never known about himself. But that cannot be, Rupert thought. Certainly, I know myself better than this stranger does, but what is he getting at?

"Lieutenant, I am no theologian; I do not study why God told us not to do those things. What I study is why *you* cannot tell me what you would do if He had never given us those commandments."

Rupert was left confused, but he understood the doctor's profession was more profound than he originally thought. For the first time Rupert looked into the doctor's face. He really looked, trying to see behind the title and clipboard to figure out who this person truly was. Rupert was struck by the feeling that this man was a great figure of authority. He seemed humble enough but there was an air of confidence about him that Rupert recognized as the unique confidence of a great leader.

Dr. Jung continued, "Now, why don't you tell me everything you remember about that dream that woke everyone up last night?"

Rupert felt trapped. There was no denying he had nightmares but considering that vision, he didn't want some doctor poking around his innermost thoughts. He decided to tell the truth about the dream but was careful not to say anything about the vision.

"I was dreaming of the last battle I flew in, back in France, or maybe over Belgium. I was shot down there."

"Ah, I had a feeling this might be the case. How often do you have this dream?"

Somehow the doctor already knew; he didn't ask if Rupert ever had the dream before, he asked how often, right off the bat. "I'm not sure, honestly. I think I have a similar dream at least once a week."

"Hmm, and is it accurate as to what actually happened?"

"Well, it starts out that way," Rupert thought through the dream in his mind. "But this time the dream was different. There was this raven that haunted me in the dream, once we get close to our target William, uh, William was my observer who sat right behind me, he disappeared and was replaced by some little girl who repeats the line, 'William is okay.' I don't recognize the girl." Rupert added quickly.

Dr. Jung quickly scribbled notes in his book. Rupert leaned in to see what they said but he was writing in what looked like German, or perhaps even Latin. He was able to decipher one word that seemed to title a whole section: *Anima.*

"What happened to William in real life?"

"He was killed in the battle; I think he died before the plane crashed." Rupert explained.

"And you were close to him?"

"He was like a brother to me." Rupert felt totally numb, which was the preferred alternative to being overcome by guilt, knowing he was responsible for William's death. The thought of all his subordinates who died came rushing back and Rupert had to fight to control his painful emotions.

"The little girl, you said you do not know who she was, but what feelings do you get when you see her?" The doctor prodded.

"I don't know, really." Rupert leaned back and thought. Whenever he saw her, he tried to look away. Seeing her was a nagging reminder that she was not William, the person who should have been in the cockpit behind him. "She doesn't really mean anything to me," he said.

"Yes, but what does she represent to you?" The doctor said in his thick accent.

This was the exact thing Rupert was trying to avoid thinking about.

"She seems familiar like a distant relative, but I don't like her. She is where William should be, and she has no business telling me what's what!" Rupert spoke louder and his voice cracked. He quickly choked back his tears and stopped himself from sobbing as he otherwise, may very well have done. "But it was all just an illogical dream. It doesn't mean anything; it won't bring William back." Rupert finished.

Dr. Jung didn't say anything, he just kept scribbling in his notes. "I want you to come speak to me later today, in my office, if you will."

Rupert did not want to go talk to the doctor but after making a scene on the hospital floor he felt like he had no choice. He felt like a prisoner again, being forced to do something he didn't want to do. Rupert took a deep breath and thought about it. He was no longer a prisoner and talking to the doctor wasn't going to hurt him.

Late that afternoon Sabina showed up to escort Rupert to the office. For a field office, it was rather nice. He had some sort of diploma hanging behind him. Trunks filled with books and other curious odds and ends were littered around the room giving the office a bohemian feel. There was a small bookshelf behind a sturdy looking desk and there was even a rug in the middle of the room.

The desk was cluttered with papers and there, lying on top of everything, was a large red, leather-bound book. Rupert was frozen in time and space, but Sabina continued to push him in the wheelchair. With an adrenaline-filled déjà vu, his eyes grew wide as he looked around the office again and realized this was the place he visited during that vision. Was the vision real? How could it have been real? Rupert thought to himself. Have I gone mad?

There was a small table with two chairs off to the side where Rupert thought to sit but Dr. Jung gestured for him to go to the sofa in the middle of the room.

"Please get comfortable. Sabina, help him to the couch."

"Oh, that's not necessary, I can just stay in my chair." Rupert resisted.

"I insist that you are as comfortable as possible, please." He said gesturing again to the couch.

I don't think I will be comfortable anywhere I sit, Rupert thought to himself, but with Sabina's help he reluctantly moved to the couch anyways.

"Thanks." Rupert said flatly to Sabina.

"Sabina is the best nurse, isn't she?" Dr. Jung said.

Sabina bashfully looked at the floor while hiding a smile. She had secretly admired Dr. Jung ever since she arrived at the hospital. The doctor with the thin mustache was handsome but his intelligence is what she really found attractive. She first met him when he was just a professor but now he was the commandant of the hospital, in charge of the whole outpost.

Dr. Jung pulled up a chair, sitting on the opposite side of the rug, and began his session.

"Do you know why you are having these nightmares?" Dr. Jung started.

"Well, I've been having dreams of the battles I fought in. I suppose I am just remembering." Rupert said blankly.

"Perhaps your mind is trying to tell you that you need to address something that happened during the battle, or battles, did you say?" The doctor suggested.

"How could my mind be trying to tell me something? My mind is me."

"If you know every bit of your mind then please tell me, why is your brain showing you that dream over and over again?" The doctor challenged.

"Once I'm back home I won't suffer these dreams anymore!" Rupert reasoned.

"I believe you would. What if you get home and you still have the nightmares?" Dr. Jung challenged. "What if there is something you need to address?"

"Everything that happened is in the past and we cannot change the past. The dream is nonsense. I just don't see the logic in talking about it." Rupert defended.

Dr. Jung paused for a few moments to think, then said, "What is the purpose of dreams? Or I should ask, if dreams are illogical nonsense, why do you keep having this same one?"

Rupert thought for a second then said, "I suppose my mind is trying to help me survive. On three occasions I have been shot at and almost killed so my brain is trying to prepare me in case it happens again so I can survive that as well."

The doctor noted Rupert's highly logical explanation but then got lost in his own thoughts about how to get Rupert to understand the imagery in dreams as symbolism. I must find a way to break through his logic so he can understand what the symbols represent about his life, Dr. Jung thought intently. He was pulled from his thoughts by a tapping noise that seemed to be coming from the window, it appeared to be a bug, so he ignored it.

"I will tell you, my boy, your mind is more than just your conscious thoughts. Tell me, what was so illogical about your dream?"

"The whole thing!" Rupert said frustrated. "William disappeared into thin air and a girl appeared, she was totally unaffected by the wind or the motion of the plane. Then she tried to hand me some sort of beetle, what's that have to do with anything?" Rupert challenged.

The tapping at the window continued, "What kind of beetle?" Dr. Jung asked curiously as he walked to the window to see what was causing the relentless rapping noise.

"It was like a scarab beetle, like what I've seen in Egypt or something. It had a green and golden shine to it." Rupert said recalling his dream.

Dr. Jung opened the window and in flew a large, shiny, green and gold beetle. He caught it and handed it to Rupert. "Is this your beetle, sir?" [FACT 25]

"YES!" Rupert exclaimed, "That is very much like the beetle I saw!" He was beside himself. "What an odd coincidence!"

"I have never seen a beetle like this in Switzerland!" Dr. Jung said excitedly. "Ha! I'm not sure what caused this but somehow, I am sure it is for you!"

Rupert's mind raced, he closed his eyes and shook his head, lost in the memory of the little girl in his dream. How could my dream have predicted this moment? He thought. The vision replayed in his mind but this time things went in reverse. He saw his body lying on the bed, then his grandfather in the armchair by the fireplace in the castle, and finally he was back in the small clearing in the secret garden. It all flashed through his mind then disappeared.

"I suppose you're right." Rupert said plainly as he snapped back to reality. His defenses were blown away and before he could think, he blurted out, "Do you know how I knew you gave me a blood transfusion, Doctor?"

Dr. Jung shook his head, no.

"I saw it happening, like a vision. It was like I was floating above my own dead body. I heard you call it a blood transfusion in German."

"Do you speak German?"

"No, but I understood anyway. Then you stepped back into the corner of the room to let the staff work." Rupert said calmly, feeling like every word he spoke reduced the weight the secret he had been carrying. "I think I can prove the vision was real." He shot a look to the red leather book that was closed on Dr. Jung's desk.

"I am very curious what else you know." The doctor said.

"That book, I have never seen it, right? After all, I've never been in this office before." Rupert asked.

"Well, I closed the book before you arrived; it is a work in progress you see. I don't think you or anyone else has ever seen inside of it."

"Well, the night I got the blood transfusion you had it open. I saw a large fancy letter D starting the first sentence on the page. It was inside a square with many peculiar images. There was a castle inside the D, surrounded by a landscape that spread into the heavens above. There was a snake that was stretched upwards. Its tail was engulfed in flames and a crown was upon its head. Is that right?" Rupert asked already knowing the answer.

Dr. Jung was shaken. He opened his book and found the page Rupert described, showing it to Sabina who was shocked at the accuracy of his description.

"I believe I was illustrating that picture on the day you got your transfusion." Dr. Jung said in awe.

"Perhaps what is most strange of all is that before I saw this book, I visited that very castle."

Dr. Jung looked at Sabina with intrigue. They sat and intently listened to every detail of Rupert's vision. All the feelings, every object, the door with God's light and how he saw himself lying on the table.

"I even saw you helping with the blood transfusion." Rupert said to Sabina. "I've never experienced anything like that, but I bet you've seen this all before."

The doctor franticly took notes, pausing for only a moment to glance at Sabina or organize his thoughts before committing them to paper. Rupert expected the doctor to say something encouraging.

"I am very intrigued with your story but, no, I have never heard of this kind of…" He paused to find the right word, "an experience quite like this."

Rupert's newfound comfort in confiding in the doctor was shattered. He felt more vulnerable than ever. He remembered all the reasons he didn't tell anyone about that vision.

"You think I'm crazy, don't you?" Rupert said, feeling disparaged.

"No, my boy! I cannot explain the things you saw, but that does not mean you are crazy." He held up a finger as he spoke, to show authority. "Likewise, I cannot explain how that beetle flew in the window at the exact moment you described it and yet it happened. You described the illustration in my book perfectly just as you described the scene of your blood transfusion correctly. It is possible someone told you about the transfusion, you may have broken into my office to see my book, and you could have made up the rest. But you only have one leg which you are still learning how to use which would make it nearly impossible for you to come here by yourself. Not to mention nothing was taken so why would you have come here in the first place? Just to tell me this grand story? If that was the case… well, I only understand the train station."

Rupert already felt vulnerable and confused but now the doctor was talking crazy. He looked to Sabina and felt a little better because she looked just as confused as he was.

Jung looked at their expressions, "Nur bahnhof verstehen," he said and looked around again. He waived his hands at their confusion and stood up to walk around the room, saying, "it just means I don't understand."

Rupert was comforted by the doctor's words but still felt vulnerable since this was the first time he told anyone about the vision. The doctor confirmed the scene of the blood transfusion had actually happened just as Rupert saw it, and the page in the book was real too. How could that be?

Rupert thought. He looked to Sabina who had been listening to the conversation. She was staring at him, wide-eyed, in awe of his story. They locked eyes and Rupert realized she was taking notes as well.

After Rupert's counseling, as Sabina pushed his wheelchair back to the hospital underneath the beautiful sky of the setting sun, he said, "You're more than just an average nurse, aren't you?"

"I am surprised it took you so long to realize. I am studying under Dr. Jung. I met him by accident and we started talking about human nature. He offered me a job, and now I am here."

"You become more interesting every day." Rupert said calmly.

"You are the interesting one. You met your dead grandfather and felt the light of God. I remember seeing you on the table; I thought you were dead." Sabina said, giving a shudder at the thought.

"I think I was dead." Rupert replied nonchalantly.

What if he *was* dead? Sabina thought to herself. "Well, I'm glad you chose to come back to your body."

For the first time Rupert felt a sense of comfort while thinking about the vision or death-journey or whatever it had been. I guess telling them was the right thing to do.

FACTS

23.   When psychiatric practices first emerged in society, many people felt as though the doctors would somehow trick them or expose their deepest darkest secrets. This is no longer the sentiment and therapy is considered the most fundamental mental health practice in our society today. [No reference required; everyone knows this.]

24.   During WWI, Dr. Carl Jung was conscripted for service as an army doctor. He was assigned as the head of a POW camp which, by necessity, also acted in the capacity of a hospital. The camp Dr. Jung was responsible for housed British troops. Jung encouraged everyone he could to seek higher education after they returned home. [Ref 20]

25.   The incident with the beetle is a true story but not for Leefe. Jung did not name the patient this happened to but described a woman who was so logical she could not see any sense in trying to decipher her dreams. She dreamed of a scarab beetle and as she described this dream to Jung, a beetle resembling a scarab actually did tap on his 3rd floor window until he let it in and caught it. The beetle did not appear to be native to Switzerland. This is an example of synchronicity and possibly the original inspiration of Jung conceptualizing the idea of synchronicities. [Ref 21]

     Synchronicity is when two events are somehow connected yet they are not caused by the same thing nor are their individual causes related to each other. The patient had a dream of a beetle and the only time a beetle ever knocked on Jung's window was when he was with her, as she described that very dream. The dream did not cause the beetle and the beetle happened after the dream and therefore could not have contributed to it. There were many factors that caused a beetle to knock on Jung's window: the wind, the temperature, the beetle's instincts, etc.

     None of those factors had anything to do with the dream or vice versa. The crux of a synchronicity is that it is the observer who *is* the connection. Without the observer to notice and connect the events, they were both just unrelated incidents… dust in the wind that require no explanation at all. [Ref 22]

Colby K.   History, Repeating Itself

Chapter 11

Victims in the Dark

The thought of Rupert actually experiencing some form of life after death and living to tell the tale was quite intriguing. It seemed mystical and bordered on the occult but Sabina had heard Dr. Jung speak of séances and other strange things before. Sabina got Rupert settled for the night and wanted to go straight back to Dr. Jung's office to discuss the revelations of Rupert's counseling. As usual other patients began calling out to her.

She hurried to provide a blanket or get some water but everyone wanted her attention for one trivial thing or another. Sabina started to lose hope of making it back to the Doctor's office when suddenly, the door swung open. Dr. Jung stepped in and in a low and commanding voice he said, "that's enough. You may go Nurse."

He commanded respect whenever he walked in a room. Everyone respected him and his authority gave Sabina a warm sense of security that felt safe. She was not like the other nurses, and he made her feel special.

Given this opportunity she immediately left the floor and practically ran back to the office where she started pondering the realizations of Rupert's life-after-death experience. Dr. Jung gave a patient a new cup of water and another a new pillow and all the other insignificant requests were immediately silenced.

When he got back to his office there was electricity in the air as Sabina spoke openly with Dr. Jung about the wildest theories she could imagine. Jung built on her theories with some of his own.

"When I was just a boy, my mother hosted gatherings where they would go into a trance to communicate with the dead." Jung explained. [FACT 26] "It would seem he did just that!"

"What if his experience is real?" Sabina pondered aloud, "*We* could document the first scientific account of life after death. How could he have known those things?"

The pair paced around the office, opposite each other, circling the couch and desk. They moved as if they were in a dance, each step was matched by the others from the opposite side of the room. They gradually inched toward the center of the room as if they were being pulled together like magnets. Their imaginations ran wild.

"He was totally unconscious, if not actually dead!" Jung agreed.

"Yet he has accurate memories from some ethereal perspective." Sabina ranted.

"Yes." Dr. Jung said. "This could open up a whole new branch of psychology!"

"It could bridge the spiritual and religious institutions of the world with actual science." Sabina realized, "We must document his whole experience; every detail!"

"We'll call it parapsychology." Jung exclaimed as he continued towards the center of the room, deviating from his circular path to move around the sofa. "There must be others! This war is not over. There will be others who have similar experiences."

Darkness had fallen outside. The office was dimly lit by two oil lamps whose light softly danced on their faces. Finally, Dr. Jung met Sabina in the center of the room. She stopped to think, but he continued circling her like a lion stalking his prey. She felt his breath on her neck. Goosebumps pricked up on her arms.

"We could unlock the next great discovery that revolutionizes the world; you and I." Jung said. The Swiss winter was cold, but the cozy office felt hotter than it had ever felt before.

His hand gently touched her back as he glided around her. "We'll do it together," he said.

She closed her eyes and felt like she was floating away. When she was with Dr. Jung she felt like she could accomplish the impossible, anything she wanted. He caressed her face and then she felt his lips press against hers. She kissed him back and then put her hand on his hip and pushed him away. She turned with flair, spinning her long hair like a ballerina's dress. She walked to the sofa and reclined on it. As she walked, she must have unbuttoned her blouse because when she laid down, it fell open, revealing her voluptuous nature.

Jung stared longingly at her, she is the most beautifully exotic woman, he thought. Jung was lost in the passion of the moment, exhilarated by the monumental advances they might provide to academia. Whether it was the vast space and time between him and his wife or the intense desire of his beautiful assistant, Jung was not sure. Whatever it was, he allowed himself to be swept away by it. The pair indulged in the great passion of the moment.

The early morning dew cooled the passions of the night before. Sabina awkwardly straightened her hair and nurse's uniform hoping to leave Dr. Jung's office before anyone saw her there at such an odd time.

Jung thought back to the night before and all the joy he felt living in the moment. It made him feel truly young again but now he was filled with shame and regret. He thought of his wife back home. What got into me, he agonized. This cannot happen again. She is my colleague and direct subordinate; it is totally improper! We must continue our work together; that pilot's experiences could lead to groundbreaking achievements. This work could change everything we know.

That very day, a courier arrived to deliver and pick up the mail. As always, there was more outgoing mail than incoming because the wounded troops wanted to write to family and friends explaining their unique circumstances. There was, however, one incoming letter addressed to Dr. Carl G. Jung.

Jung recognized the handwriting at once and wondered if this was some cruel joke God was playing on him. His morning began with guilty feelings for his extra-marital relations and now he was faced with such a coincidence that it all seemed preordained. The letter was from his wife. She described her loneliness and how she longed for his return, but it also seemed like a dire warning.

> *Dearest Carl,*
>
> *These days I am all too familiar with the loneliness of a cold bed. I miss your warmth dearly and dream of the day we are together again. I am keeping busy with my personal studies. I think it dreadfully unfair that I did not receive a proper education but I suppose whatever life does not do for you, you must do for yourself.*
>
> *No matter how much I focus, I cannot help this nagging thought that continues to irritate me to the point of distraction from my daily life. I worry that you will return to me a different man, one who is not true to me and the vows he took at our matrimony. I understand the drives caused by human nature but I pray you will be able to abstain from your carnal desires until we are reunited.*
>
> *I will never be jealous of the time you give to your work but I must retain my rightful place in your life. I do hope you will write to me soon and quell the voice in my head that distracts me. I have a vision of a day in the future, after this war is over, when we are able to share our lives again, as we once did.*
>
> *As Always, With Love,*
>
> *Elizabeth*

I do admire her. There is no choice for me, Jung thought to himself. I must remain with my wife. It was love at first sight almost 20 years ago. She is the foundation of my home and her inheritance funds my life's work. What are the odds that I receive this today? I am in the midst of deciding how to study the hidden nature of coincidences, and I receive this letter only to experience one myself. Perhaps modern science could give some explanation as to how these totally unconnected events may actually be connected. On the other hand, the Bible describes God as unfathomable for our minds to conceive. Perhaps we will never understand.

A few months later,

"I had the dream again." Rupert reported, as he lay in the reclining chair, once again in Dr. Jung's office. "This time was a little different. William turned into the little girl again but she didn't offer me the scarab beetle. Instead, she was clutching a white scarf- it was something we wore as part of our uniform. This one was soaked in blood and somehow, I understood it was William's."

"Interesting," was all the doctor said. He wrote many notes. It seemed that he wrote just as much, if not more, than Rupert actually spoke. "I believe the subconscious mind uses symbols in our dreams to tell us things. I just have to figure out what the symbols mean."

"So that's it, then?" Rupert said, feeling a bit like a lab rat being studied and less like a patient receiving treatment.

"There is one thing I must tell you before we end this week's session. I believe, at this time, you are healthy enough to be released from this hospital. You are ready to be sent home." Jung told Rupert.

"Home." Rupert said, sounding disappointed.

"Now, I believe we still have some things to discuss, and I invite you to stay so we can explore more of your dreams, but I must leave the choice to you."

Rupert thought for a few moments. He thought of the family waiting for him and what he would do if he returned. Sabina was not present for this session, but he thought of her. He had no sweetheart to run home to. There was a knock on the door and Sabina walked in to whisper something to the Doctor. Rupert stared at her. He felt something he couldn't explain but he knew it was something he should stay for.

"I'll stay here, Doctor." Rupert said confidently, embracing his newfound place in the world.

Sabina looked at him and smiled. She helped him stand up, aided by his crutch and escorted him back to the hospital.

"What made you decide to stay?" Sabina asked.

Rupert felt like he had nothing to lose, "Honestly, it was you."

Sabina blushed and looked at the ground but didn't say anything.

"Sorry," Rupert said quickly. "I know it isn't exactly appropriate for a nurse and a patient to be involved."

"Yes, it is not appropriate." She said sounding disappointed before looking up to Rupert. "But sometimes we cannot control how we feel." She gave a halfhearted smile and moved her hand from supporting his arm to holding his hand.

"Stay with me." Rupert said with a subtle but confident smile that Sabina was instantly attracted to. "Let's go for a stroll, this weather is almost as beautiful as you are. Come on, I think I'm strong enough to make it the whole way into town."

"Perhaps one day we will, but right now it's my turn to talk to the Doctor." Sabina said, with a touch of disappointment in her voice.

"Wait, you go for counseling as well?"

"Yes. It's part of my training as his apprentice. I will see firsthand what the patient goes through while learning about the different layers of consciousness." Sabina explained. "Dr. Jung told me that receiving counseling has been a personal weakness for him and he wants better for his pupils."

"For the longest time I thought you were just an average nurse. Now I can see that you are so much more." The pair reached the hospital door. "Until our next rendezvous." Rupert said before watching Sabina walk back to Dr. Jung's office.

"Before we begin there is something I must address." Dr. Jung said to Sabina. "It was wrong of me to begin such a personal relationship with you while you are my understudy."

"We haven't spoken of that night since it happened," Sabina said softly. "It is okay Doctor, I wanted it too."

"This is different. It would be even more inappropriate to continue a relationship of that nature during a period of counseling." Jung paused for a moment. "You must be able to confide in me without having me as a subject of your unconscious emotions. Can we move forward only as colleagues?"

Victims in the Dark

Sabina thought for a moment. She admired Dr. Jung. He was handsome and intelligent. She wasn't sure she could simply forget the connection they shared or quell her very real romantic feelings towards him. She had to tell him something. If she said no, then perhaps they could remain lovers instead of colleagues. But that would be a risk, she reasoned, he might cut me off altogether; after all, he is married.

"Yes. I understand your position. I want to continue working with you." She relented.

"Very good." Dr. Jung said with a sigh of relief. "Now, while the pilot was speaking, you mentioned being distracted by something from your past." Dr. Jung solicited.

"Yes. Something happened that I have not fully processed. It is like a dark shadow lurking in the back corner my mind and I fear that when I provide counseling for patients it will prevent me from being helpful." Sabina explained.

"Please tell me what is bothering you, what happened?"

"I was out with my friend, a neighbor I grew up with. We had just enjoyed an opera and she had been invited to a party. That is where I met him." Sabina dove into her memory.

"Frederick. I noticed him immediately because he walked with a strange limp. They told me he had been injured in battle and when I looked into his face, I saw what they meant. He had a large scar across his right cheek. A little while later, my friend left with the man she was dating and I knew I would be walking home alone that night. It was only after she left that Frederick approached me.

"It's a cold night for you to be alone." Frederick said in his distinct raspy voice. I don't remember what I said but I told him I was just leaving. As I walked home, I got to a corner where I had a choice. I could stay on the well-traveled roads but it would take an extra 10 minutes or so, or I could cut through the alleys and get home quicker. My feet were hurting from the shoes and my corset was tight and uncomfortable, so I chose the quicker route.

I never expected that choice would change my life so dramatically. I reached the second turn and a dark figure jumped out of a shadow and grabbed me. I had heard stories of men who do bad things but I never thought I would meet one. All I saw was the scar on his face before he hit me. He used me until he was satisfied. I felt so powerless. The whole thing was awful but his horrible, raspy laugh was so vulgar and so horrid that it still bothers me. I opened my eyes as he limped away. I wasn't completely sure who he was until that moment. The limp, the scar, and the raspy voice; it was Frederick from the party."

Sabina shuddered as she spoke, her arms folded tightly. She wasn't shaking or anxious. It was like she was numb to the whole thing.

"I know this is difficult; the first time you talk about it can be like experiencing it all over again." Dr. Jung said. "You must discuss it, if you want to move on with your life, and you are doing very well."

"Doctor, I wish to love again. Like I used to love before this happened." Sabina longingly stared into Dr. Jung's face as tears began to swell in her big round eyes. "I think I will, with your help. Only you understand my thoughts and feelings and only you can help me love again!" Sabina pleaded, leaning towards him.

"My dear Sabina. I cannot love you in that way. It is against all my ethics as a doctor." Jung replied sorrowfully.

"Do you deny that we have a connection? Do you deny that our love is true in both mind and body?" Sabina challenged.

Jung sighed. "I deny nothing. But right now, what you need is not my affection but your own acceptance. I must retire for the evening. It has been a long day for both of us." Jung instructed as he stood up and walked behind his desk.

Sabina was left feeling vulnerable. She wasn't exactly rejected but that is exactly how she felt. She dried her silent tears and found solace in returning to the hospital floor where she could help people.

Whenever the patients called out for her, she felt needed, she felt desired again, even though it felt superficial. In the back of her mind, she kept her dark past hidden from them all. She felt like if they knew what happened they would be repulsed. Still, she enjoyed stopping every so often to check on each patient, giving them fresh water or changing their sheets. Anything to put the past, and her rejection from Dr. Jung, out of her mind.

Victims in the Dark

She came to Rupert's bed. He just looked at her and smiled. She felt warm inside, like he accepted her for who she truly was, even though there was much he didn't know.

"How about that stroll around the grounds?" Rupert said with a smirk.

Sabina smiled, "Yes, I think that would be nice."

The sun had disappeared behind one of the tall mountains even though there was still plenty of daylight left in the evening. The moon was peaking up over the top of a snowcapped peak in the early twilight sky and the air had already cooled considerably.

"This really is a wonderful place." Rupert said admiring the beauty.

Rupert managed to transition from his crutches to a bench, sitting in the long shadow of the setting sun behind the mountains. Sabina sat next to him, as the cold breeze blew over them. Rupert had a newfound confidence with the realization that he truly was not a prisoner. For the first time, perhaps ever, he was really choosing to be where he was.

Rupert quickly took the opportunity to comment on the cold and slid his hand over her shoulder, landing his arm around her. Sabina snuggled into his comforting embrace and the two sat on the bench in silence just watching the day fade until they were surrounded by twilight.

"This is nothing like the place where I grew up." Rupert said, finally breaking the silence.

"What was it like there?"

"When I was a young boy, I lived in the jungles of South India. My parents ran a coffee farm; it was nice there, very warm. But my home? That's a place in England, a little town right on the sea." [FACT 27]

"My home is also a city by the sea. During the summer we would go out to the beach and play in the surf and the sand." Sabina reminisced about a simpler time. [FACT 28]

"I loved to sit in the grass up the hill from the beach and watch ospreys diving for fish in the water." Rupert said.

"It seems we have more in common than we knew." Sabina said softly as the pair enjoyed each other's company until night fell.

FACTS

26.   Dr. Jung's mother hosted séances that young Carl was witness to and participating in. The seances were specifically for the purpose of contacting the spirits of people who passed. This topic will be addressed in later Chapters. [Ref 23]

27.   Leefe Robinson was born and raised in India where his parents owned a coffee farm. He lived there until around six years old when he moved to England. [Ref 24]

28.   The character Sabina Sklyar is based on the real-life Sabina Spielrein. She was from a city in Russia called Rostov on Dom that is a few miles from the Sea of Azov, North of the Black Sea. [Ref 25]

Chapter 12

A Warm Wind Blows

In August 1918, Dr. Jung was summoned to Zurich which abruptly halted both Rupert's and Sabina's therapies. Dr. Jung hoped his absence would cool Sabina's passions and they would be able to continue their work on Rupert's experience in the afterlife. Weeks passed and it was autumn before Sabina found herself back in Dr. Jung's office picking up therapy where they left off.

"After it happened, I felt so many things." She explained. "Anger, pity, sorrow, pain. I found my way to a public place and someone summoned the police who took me to a hospital but I don't remember much of that. The next day I was with my friend, her lover and his friend and I told them what happened. He called for Frederick who showed up but, of course, he denied it. They looked to me and I described my attacker. I told of the laugh and the scar on his face, and he defended himself saying that many men had scars just like his.

Frederick claimed he went home alone that night. Someone confirmed he was seen leaving just after I did. They weren't convinced until I described the way my attacker limped away. He called me a liar and raised his hand to strike me. My friend's boyfriend defended me and a fight broke out. He threw a punch but Frederick landed one first. The friend hit Frederick with a piece of wood which he must have grabbed from the pile by the fireplace. He fell down hard and hit his head. I will never forget the sound it made as it hit the stone floor. Frederick ended up dying sometime later because of that."

Dr. Jung listened to her story but offered little advice. Perhaps he was still treading lightly to avoid stoking the passions they were both trying to suppress.

"There is something else, Doctor." Sabina said. "I've been having a dream."

"You think it is related to this?" Dr. Jung asked.

"It is this." Sabina said dramatically. "I'm at the party before it happened and I lock eyes with Frederick. As I stare at him, he transforms into a raven."

Dr. Jung let out a faint gasp but Sabina continued.

"In the next moment I am in the alley. He is still a raven but somehow he is transforming me into a raven. I don't understand what it means."

Dr. Jung asked a few more questions but they came to no conclusions. Sabina still longed for Dr. Jung's acceptance. Even though she was reliving the worst thing that ever happened to her, she still wanted to feel desired. Instead, she felt more alone than ever.

Sabina left Jung's office feeling vulnerable. Logically she knew she would never be with Dr. Jung but the heart does not abide by logic. She walked straight to her living quarters so she could be alone for the rest of the evening. As she walked down the path, she was quite surprised to see Rupert on his crutch.

"Where are you going?" She demanded.

"I'm going to find a pub in the village and there is nothing you can do to stop me." He said, proud and defiant, like a rebellious teenager.

Sabina looked at him like a concerned mother, which is how she often looked at her patients, but her concern melted into a playful smile. "Then perhaps I will join you!" She said with a laugh.

A Warm Wind Blows

Rupert was elated and the pair found their way to a pub. After a few drinks they were the life of the party. With each other, in that moment, they forgot all their problems. They stayed late into the night until Sabina convinced Rupert they would be missed at the hospital.

The pair stumbled out of the bar, Rupert clinging to his crutch. Given their current state, they were not going to make it one hundred yards, let alone the entire way back up the hill to the hospital. As they lumbered on, Rupert's crutch caught a snag in the rocky street and he tumbled into the grass. Sabina reached out to catch him and ended up falling with him. He hit the soft ground and started laughing hysterically. She laughed playfully as she accidentally rolled directly onto him until they were face to face.

Their laughter dwindled into smiles as they felt each other's warmth. Rupert kissed Sabina with all the passion he had bottled up over the months since he first laid eyes on her. They stumbled to a guest house where their feelings combusted into a blossoming consummation of drunken love. They spent the night embracing their passion for one another.

A few days later, Jung summoned Rupert for counseling.

"I'd like to describe an idea to you and see what you think." Dr. Jung began. "When you went into battle and entered that blissful state you described, what caused you to forget your natural fear of death?"

"I have done some thinking on this," said Rupert. "I believe I was so certain I was going to die that I just let go of the concern for it. Of course I feared death, but in all my actions and thoughts it simply wasn't a consideration. Normally I wouldn't have taken risks that would have put me in danger but since I felt like I was already going to die the risks didn't add any more danger. I was free to act with no regard for my safety."

"You accepted your death as imminent, so every moment was extra time." Dr. Jung thought aloud. "I think that you...you entered a state of mind in which you were both alive and already dead. Of course you were alive, because you continued to exist but the odd thing was you embraced death and that is why you were able to live."

"I suppose that is accurate," said Rupert. "I'm not sure I fully understand what you are trying to suggest as far as being both alive and dead at the same time. I assure you, at no point was I dead." As Rupert spoke these words, a vision flashed through his mind. He was in the great room inside the castle, sitting with his grandfather, then he saw himself hovering over his body in the hospital bed causing him to regret his last comment.

"I received some correspondence from my English colleague, Peter Baynes. They are calling it 'shell shock.' People who experienced battle are suffering night terrors, just as you have. Many go into fits of uncontrollable shaking, even while they are wide awake." Jung explained.

"Well. I suppose I'm glad to hear I don't have it as bad as it could be." Rupert tried to sound optimistic.

"There are others who are having the same reaction you are. That tells us you are perfectly normal!" Dr. Jung said jovially.

Rupert felt a sense of acceptance he had not had since he was back at Sutton's Farm. After a moment he said, "You know doctor, I must admit, while I was having that odd vision; perhaps I was dead without being fully dead."

"Yes, I have thought of this possibility." Dr. Jung continued, "To tell you the truth it is what made me think that somehow you achieved that state during the battle. When you were flying, I don't mean to suggest that you were actually dead. Rather your mind convinced you to forget about life, objectively. The impending doom was replaced by clarity which ultimately saved your life."

"My brain tricked itself into relaxing." Rupert stated, as he thought about it.

"Not your brain, your mind." Jung quickly corrected. "I believe we may be able to treat this shell shock by bringing patients into a state of living-death." He appeared to stop talking to Rupert and was just thinking out loud. "The question is how do you get someone to explore the inevitability of their own death without actually endangering them or exposing them to an unethical amount of stress?" [FACT 29]

"Doctor, I'm not sure about any of that but can I ask you a question?"

"Of course, my boy."

"I've been pondering something you said. When you were asking me about the ten commandments, what were you getting after?" Rupert asked.

Jung thought for a moment before saying, "Well, my point was that you have a dark side about you that you neglect. Not just you, each of us. I call it the shadow and if you ignore yours, it will eat away at you. [FACT 30] All individuals are capable of breaking any of the ten commandments. Without those learned values, you may do some of them. You may even take great pleasure in doing them."

"I must disagree with you. I would never want to do any of those things." Rupert argued.

"Hmm, what is the sixth commandment?" Jung challenged.

Rupert wasted no time as he was well versed in the commandments. "Thou shalt not kill."

"And yet you, yourself were once in a plane trying to kill people." The doctor struck a nerve.

"That is not the same! It is different in war. I wasn't trying to murder; I was trying to do my duty!"

"I am not trying to upset you," Jung said calmly. "But the truth of the matter is that the commandment says you will not kill. It does not say you shall not kill unless it is necessary for your duty."

Rupert was angry. He felt like he was being personally attacked. I am a righteous person, he thought. I am human and have my flaws but I do not outright abandon the commandments of God.

"I think we are done for today." Rupert said as he sprang up onto his crutch.

"I'm not trying to challenge your beliefs. I'm trying to find truths." Dr. Jung said in vain as Rupert left the office. [FACT 31]

That night Rupert had the worst dream yet. This time it wasn't just William; it was all the men from the Flight he commanded. They all died because he led them to their doom; and that night he was forced to relive all that anguish.

"The pilot's dreams seem to have gotten worse, Doctor." Sabina reported to Jung, alone in his office the next morning.

"Hmm, perhaps we must do something different with his counseling, but what?" Jung thought aloud. "We will figure something out for him, but for now, it is your turn. I feel there is more to your story."

Sabina did not want to talk about her experience again but she had been looking forward to this time with Dr. Jung all week. She dressed up as best she could and even found a late-summer flower to rub on her wrists, hoping it would act as a perfume.

She was still secretly hopeful they could have the relationship that once seemed to be blossoming. For the time being, she complied with his instructions and told him about a dream she had.

When Sabina told Dr. Jung the dream, he started getting excited like he was a hound dog tracking a scent.

"Is there something else that happened? Some part of the story you've left out?" Dr. Jung questioned.

"I don't think so." Sabina said.

"What happened later? Were you able to spend time with your friend after the event?" Dr. Jung asked, looking for the missing part of the story.

Sabina realized there was one last thing to tell. After hearing this missing piece of the puzzle Dr. Jung was left curious and excited. All he said for the time being was, "Fascinating."

During his days, Dr. Jung roamed the hospital overseeing operations and interacting with patients as much as he could. In the evenings he would retire to his office to contemplate the vast intricacies of the human psyche. It was during this reflection that he thought through the similarities of Sabina and Rupert's stories.

Sabina was turned into a raven in her dream and the Pilot was haunted by a raven in his. The raven could symbolize the dark events they each endured or even death itself. But how could the same symbol mean the same thing for two different individuals?

This knowledge might get passed down to us, Dr. Jung thought, or perhaps we learn the symbols through our culture.

No, that is unlikely in this case, Jung's internal dialogue retorted, Sabina is Russian and the Lieutenant is English, they would have been exposed to different stories and learned different symbols. It must be something deeper. Perhaps the knowledge of these symbols is passed down through genetics just as any instinct would be inherited.

Jung's thoughts jumped back to the dreams and what each might symbolize. Rupert's raven could symbolize death, he's seen plenty of it. I think Rupert feels personally responsible for the death of William and the rest of his men.

"Ah ha!" Jung said aloud.

His raven symbolizes the death of William and his men. Sabina seems to have a shadow very similar to Rupert's. It's possible they both have the same feelings towards their unique experiences. There is a connection, I just have to make them realize it, but how?

Perspective, Dr. Jung thought. He recalled all the times Rupert discussed his dreams involving ravens. Sabina had not been present for any of them. She was always busy with her nursing duties.

Perhaps if I counsel them together, they will each see their own story within the others'.

FACTS

29.   In modern medicine there is an experimental treatment for Post-Traumatic Stress, or PTS, that effectively makes patients forget about the fear of death. This treatment is Psilocybin. Magic mushrooms and Ayahuasca, or DMT, are potential treatments for PTS which can exhibit itself in many forms. Ultimately the mind is experiencing the fear of dying or repeating a similar experience as what caused the PTS in the first place.

These experimental drugs can be taken in micro-doses where the patient remains totally sober or given in larger amounts which can allow the patient to go through a deeply transformative, guided meditation. These substances create a deep sense of feeling connected to the universe and for some, they overcome their debilitating fear of death.

Likewise in Near-Death Experiences, individuals usually report their psychological fear of death dissipates or is totally replaced by an acceptance of this inevitable fate.

The individuals never lose their fear of physical death, that is, they still have all their instincts and natural desire to sustain life. What they lose is the debilitating fear of death that interrupts their life or even prevents them from functioning altogether. [Ref 26]

30.   Archetypes will be discussed in greater detail in future chapters.
          The Shadow Archetype - Dr. Jung conceptualized a mental block, of sorts. Hidden in the subconscious mind, we have two self-images. The Self and the Persona. When the public-facing self (Persona) and the whole-self (Self), do not match; the shadow is created.
          We all do it. The persona we believe we are, aligns with our values and never deviates from them, but as humans (imperfect as we are), our actions sometimes deviate from our values. When our true self does not align with our ideal self, the mind creates a shadow so we can go on believing we are true to our values.
          The bigger the lie, the bigger the shadow becomes. If the individual consciously addresses their shadow they can make peace and move on. Without addressing the shadow, it will continue to pop up into the person's life in dreams or even cause the person to act 'out of character.' [Ref. 23, 27]

31.   In the narrative, Rupert's Persona or public-facing self-image is that of a good Christian but Jung challenges that by making Rupert realize he broke one of God's commandments. Since a good Christian (Rupert's Persona) would not have done that but Rupert (Self) did do that, Rupert is left with a conflict between his Self and Persona. This conflict is the shadow which Jung exposed to Rupert's conscious mind.
          More explanation on the shadow and other archetypes will follow in the latter half of The Owl and the Osprey.

Chapter 13

(Post-Traumatic Stress)^2

"How is talking about this again going to cure my shell shock, Doctor?" Rupert complained from the couch in the center of Jung's field-office.

"It is time for me to... how do you say it, come clean. I think your case of shell shock was cured when you had the near-death-experience."

"How could that be? The nightmares have gotten worse since then." Said Rupert.

"The first time we spoke about your vision you said something I thought was quite intriguing. Do you remember describing how you felt when you woke up?"

"Well," Rupert began, "I remember feeling like the vision was more real than being awake. Like that was the true reality and this is the dream."

"Yes, but there was something else you said."

"How did you feel about dying after your experience?" Sabina asked from her own chair, next to the sofa.

"I remember feeling like no matter what happened during life, everything would be fine in the end."

"Yes!" Jung exclaimed. "That is exactly it! That is the feeling! That deep understanding that moves you past the fear of death and cured your shell shock."

"Wait, if I don't have shell shock then why do I keep having these awful dreams?"

"Shell shock is a physiological response, but you do not suffer from uncontrollable shaking or any other typical symptoms. I believe your night terrors are caused by something other than your traumatic experiences. Something even more personal to you." Jung said mysteriously.

Again, Rupert felt like the doctor knew more about him than he knew of himself. This was both intriguing and irritating.

"And Sabina," Dr. Jung said, turning his whole body towards her. "Do you feel you have dealt with the things that happened to you that night?"

"What do you mean, dealt with?" Sabina said abrasively.

"I'm sorry. I don't mean to patronize you." Jung turned to Rupert to fill him in. "Sabina went through a traumatic experience where her life was in jeopardy." Looking back to Sabina, he said, "My theory is that you cannot cope with what happened to you until you address something related, something that is bothering you even more."

"What could bother me more than being raped?" Sabina said dramatically. She knew Rupert would soon find out, so she just came out and said it. For the moment, she looked down rather than meet Rupert's gaze.

Rupert cared about Sabina; he may have loved her. He leaned over and slowly touched her hand, trying to give her reassurance. She recoiled.

"I do not need your pity, Rupert." She snapped. Rupert was lost, he wanted to help but had no idea what to do so he looked to Dr. Jung.

"Stop." Dr. Jung said holding up his hands, "We have already gone too far. Sabina, I have already spoken to you about this session so Lieutenant; we must first set expectations and ground rules. This is a group counseling session. You must promise not to repeat anyone else's story. Their story is not yours to tell. Do you agree?"

"Yes," Rupert said quickly.

"Of course." Sabina calmly said a moment later.

"I want both of you to be able to tell your story completely without holding back because you fear negative judgement." Jung continued. "I figure the two of you are strangers brought together by happenstance.

(Post-Traumatic Stress)²

Once Rupert returns home you will likely never see each other again, so you can each speak freely."

These words felt like a dagger in Rupert's chest. He had envisioned a future in which he and Sabina met after the war. He even entertained the idea they might be in a committed romantic relationship. He looked at Sabina who seemed to be in a daze, staring at a spot on the ground.

Sabina felt Rupert's gaze calling out to her. She sat up straight and stared into his vulnerable eyes. She cared for him too.

"How will we begin, Doctor?" Sabina asked.

"One of you must tell your story; followed by the other, then we will have some discussions based on the similarities."

I almost died in war, Rupert thought, how could that be similar to getting raped?

"I would normally offer ladies first, but here I feel its best if I lead." Rupert said with confidence. He felt it was the chivalrous thing to do.

"As you know, I was in a plane crash on my first mission." Rupert relived every bit of his mission to destroy the German artillery. He described every feeling, starting back in England with his first encounter with the zeppelin. The clarity he got when he was actually in battle, the rocky coastline he dreaded, the way he felt so desperate and helpless as he pulled William out of the cockpit before his leg gave out. The guilt he felt for dragging his mates into the war and the fact that he was personally responsible for their deaths. He told everything the whole way up to when he awoke in the hospital. Throughout his story, he often looked to Sabina who seemed to be in deep contemplation.

"Could you please describe your dreams?" Dr. Jung asked after he finished.

"I was flying in my last battle. Every direction I turned my plane I couldn't help but follow this giant raven."

Sabina gasped as if Rupert said some secret word. Dr. Jung was already watching her as if he knew a reaction was forthcoming. Rupert paused for a moment, puzzled, but then continued.

"It seemed like the raven knew I was going to turn before I did. It wasn't towing me because I was able to steer any direction I chose, but somehow, I was forced to follow this foul creature. It's like the raven knew what I was destined to do so it was one step ahead of me. At some point William vanished, as he usually does in the dreams. I kept flying to the artillery until I was so close I could see the faces of the men I was supposed to destroy. They were all William. It's like I had been sent there to kill him. I've also had dreams where the raven chases me. One time the raven sat in the cockpit with me."

"This is a very interesting development." Jung said, containing the excitement in his voice as he looked directly at Sabina.

"Why, is that?" Rupert asked.

"I will tell him." Sabina said. Rupert was quite surprised. Rupert was frustrated that everyone in the room seemed to know something he didn't and yet he had been the only one speaking for at least half an hour.

"As you heard earlier, I was raped. It was back in my home country, where I grew up. I was out with my friend at a party but we got separated and so I walked home alone."

"Wait, wait, wait." Dr. Jung interrupted. "You are not speaking in detail. What happened at the party?"

"That has nothing to do with what happened to me Doctor!" Sabina defiantly replied.

"Sabina, why is it that you feel perfectly fine talking about the worst part of it but you avoid telling other details?" Jung asked gently.

"I just don't see what that has to do with anything," she said. Jung stayed silent and gave her a look that clearly instructed her to do as he said.

"If I must." Sabina began again. "My friend fell in love with a man there. I told her I'd be fine and so, she left the party with him. Before she left, I asked her about a man I noticed from across the room. He had a large scar on his face and he walked with a distinct limp."

"Mm hm." Dr. Jung mumbled, leaning back in his seat.

"And so, I was walking home and the man with the scar on his face found me in an alley. That's where it happened." She said completing her recap.

Jung looked at her from the corner of his eye and frowned. Rupert noticed this and still couldn't figure out how his story related to hers.

Sabina caught the look Dr. Jung was giving her, so she continued, "A day or two later I was telling everyone about what happened. That's when

Frederick walked in. A fight broke out and Frederick hit his head by the fireplace which ended up killing him." She choked back tears as she finished her story, it was still hard to talk about.

"Tell him about your dream." Jung instructed.

"Yes, my dream." Sabina said gently. "You are not the only one who has dreams around here, Pilot."

Rupert smiled but felt that something was off, she had always called him by name, and never just, 'Pilot.'

"I have been having a dream where I am back at that party. I am staring at the man with the scar, we lock eyes and then he turns into a giant raven."

"Oh." Rupert was stunned. She dreams of a raven as well.

"I get lost in his black eyes," Sabina continued, "and then I am back in the alley. I don't exactly know what he is doing to me in the dream but somehow, he is turning me into a raven. When he is finished, he simply disappears and I am left as the raven itself."

She *becomes* the raven, Rupert thought, how is it possible that we dream of the same thing? His mind was ablaze with thoughts, and he did not direct them; instead, he let them run wild until it hit him.

"The raven was me!" Rupert exclaimed aloud before he realized he was speaking. Dr. Jung and Sabina just looked at him, Jung had a smile of accomplishment creeping up from the corner of his mouth. "That's how it knew which way I was going to turn. William turned into my target. It's like I was trying to kill him." Rupert stopped there before his emotions got the better of him.

"The things you see in your dreams are only symbols. What does a raven symbolize to you, Rupert?" Jung asked.

"I don't know." Rupert said initially. "A bad omen I suppose, or perhaps death itself."

"Yes, both of these things could be true. If the raven was you, as you exclaimed, and it symbolized death leading you in some way to your target, then what would the dream mean?" Jung challenged.

Rupert felt vulnerable, he didn't want to answer the question for fear of what he would realize but some part of him knew he needed to answer, not for the doctor but for himself. He pushed through the awkward feelings and said, "I suppose, if I was the raven and it was leading me to my target, which was William, then it's all a big reminder that I killed William." Rupert was getting frustrated and yet he continued to delve into this conundrum. "He actually became the target and I was the raven hunting him." Tears began running down Rupert's face as he continued. "I never wanted him to die, I never wanted to kill any of my mates but now they're all dead because of me!"

Rupert began to sob and for a few moments was inconsolable.

"No. They are not." Sabina said softly, sounding almost angelic as she spoke. Rupert looked over at her and saw she too was crying. "Rupert, you did not kill your friends. Your mind is trying to show you that you are experiencing survivor's guilt."

"Yes," agreed Dr. Jung in an authoritative voice, "it is not shell shock but survivor's guilt that torments you, my boy."

"Survivor's guilt?" Said Rupert who had never heard of such a thing.

"Doctor," Sabina said as it all clicked for her. "You are truly a genius. Rupert, you are tortured by your own, incorrect, thought that you killed your friends. You did not shoot your friend. The Germans did. All of their deaths were out of your control and you must let go of your misplaced guilt so that you may live a full life again. It is what your friends would want for you."

"Excellent, Sabina, well done." Dr. Jung said, trying to encourage her towards her own realization without pushing directly.

"I can see now," Sabina said, "the cause of the ravens in our dreams is the same. The Doctor wanted us to see our own diagnosis within each other's story." In that moment, Sabina couldn't say anything more. She was tongue-tied by her emotions and stared off into nothingness.

"Wait, I don't understand; how my raven is connected to yours." Rupert said looking at Sabina for an explanation. She stayed quiet for a moment until her voice broke the silence like shattering glass. In a way, she sounded empty, like she was totally numb.

"Maybe a month after Frederick died my friend tried to make me feel normal again, even though I didn't think that was possible."

(Post-Traumatic Stress)^2

Russia, Years Earlier

Sabina and her friend sat at a small table outside a café in a quaint city square nestled at the edge of a bustling European city.

"How is your cappuccino?" Sabina's friend said.

"It's okay." Sabina answered without emotion. The early summer flowers were in full bloom in flower beds throughout the square but to Sabina everything looked grey.

Across the plaza, maybe 30 meters away, a man sat down on a bench. Sabina's eyes haplessly wandered across the square until they froze on a jagged, ugly scar on the man's face. She felt adrenaline course through her veins but her whole body was paralyzed by fear. As if she was being tortured, her mind jumped back to the night she met that exact same scar in a dark alley.

"Are you okay?" Her friend asked, alarmed. Sabina was shaking but didn't say anything. The man caught her gaze and stared into Sabina's eyes. He recognized her and smiled.

Sabina let out a gasp as if she had seen a ghost. She couldn't speak and appeared as if she was about to faint, so her friend quickly brought her to the restroom.

"That was him!" Sabina gasped, still shaking in fear, with tears running down her face.

Her friend hugged her, hoping to calm her down.

"Sabina, Frederick is dead, you are safe here."

After a few deep breaths, Sabina calmed down.

"I know Frederick is dead, but the man out there isn't Frederick. His eyes just… they hurt me. I can't explain it!"

"Let's go back out. I will be with you, if he is still there I will…" her friend was getting excited and didn't know what she was going to say, let alone what she might actually do in that situation. "I'll kill him!" She said fiercely.

The pair walked back to the table where Sabina pointed to a now empty bench across the city square. They sat back down and Sabina took a deep breath. Maybe it was all in my head, Sabina thought.

"I don't see our waitress. I'll go in and pay then we can get out of here." Her friend said, leaving Sabina at the table.

Sabina nervously turned to look inside the café at her friend standing at the counter trying to pay the distracted barista. When Sabina turned around the man was sitting at her table. She looked into his horrible face. This man had a scar almost identical to Frederick's, though his face was rounder. She was paralyzed with fear like never before.

"My name is Valdrom, I never got yours." Her assailant said in a raspy voice as the corner of his mouth turned upwards in an evil smirk. He had wild eyes that sent chills down her spine. "Oh, come on, I had fun with you. Now you show up here toying with me."

Sabina felt like she was going to vomit. She was petrified but the look on her face showed her disgust. She was no longer shaking and while she thought she should be scared, she felt nothing at all.

"I'll be out again tonight, waiting for you." The man chuckled that same horrid laugh that Sabina had replayed in her mind a thousand times. Valdrom stood up to walk away without any hint of a limp.

I remember him limping. She thought. It was so clear that night; or was it? I had just been attacked, maybe I couldn't see properly. But I accused Frederick *because* of the limp. He is dead because of me, what did I do?

"Who was that man?" Sabina's friend panted as she sprinted back to the table after seeing the stranger walking away.

Sabina looked at her friend with tears in her eyes. She said nothing which told her friend everything.

Switzerland, Present Day

Sabina faded back into reality in the comfortable office nestled in the foothills of the Swiss Alps. "I suppose the doctor already figured it out. I cannot fully cope with being raped until I address the fact that I am responsible for the death of an innocent man. I wish I could make it up to him or those who cared for him."

"I have found that leading the patient to their own realizations is more effective than telling them." Jung explained, "They get a more personal sense of understanding that way."

"Doctor, you are a genius!" Sabina said enthusiastically, reaching out and holding one of his hands with both of hers, "You are such an incredible mentor."

Rupert couldn't help but feel uncomfortable with how passionately Sabina was speaking to Dr. Jung. Rupert had been harboring serious romantic feelings for her and even consummated them, but he was quickly realizing that Sabina was infatuated with the Doctor.

"I only mean to help." Jung tried to keep the conversation professional as he pulled his hand away, leaning back in his chair.

Sabina felt exposed. She just aired her darkest memories and now had nothing to lose. She was so focused on Dr. Jung that she seemed to forget Rupert was there at all. She leaned over to Dr. Jung and took his hand in hers again. This time he recoiled and stood up to walk behind his desk, not wanting to exacerbate an already volatile relationship, especially in present company.

Rupert saw this and felt betrayed. He wanted to lash out at Jung, but he seemed to want nothing to do with her. What is going on? Rupert thought.

Sabina stood up and followed Jung behind his desk, speaking vehemently. "We have a special connection; you are so much more than a lover to me, Doctor!"

Jung continued to walk around the room. Sabina followed, slightly hunched over, shuffling her feet along the floor while leaning on the furniture as if her hands were helping her walk.

"Do not deny it! You and I have something special. We have a mental connection even greater than our physical one!" She continued.

Jung kept moving away from Sabina, trying to keep some piece of furniture between them as if she was a hunter and he was her prey. He said nothing leaving Rupert's mind racing. What did she mean physical connection? He dreaded what the answer might be. The thought of the bed they shared in the Swiss village was still fresh on Rupert's mind, but now she is professing her love to another man. How could that be?

"STOP RUNNING AWAY FROM ME!" She screamed. Jung stopped. He realized he had to address her feelings, even in front of Rupert. He planted his feet and stood up tall.

In the moment, Sabina felt more vulnerable than ever in her life. She was emotionally naked, having told her deepest darkest secrets, inadvertently, to both of her romantic affections. She saw the look on Rupert's face and she knew he would never look at her the same. It turned into the perfect storm. She half-wondered if Jung had orchestrated the whole thing to ensure her love life would be totally ruined. I may never love again, she catastrophized.

"You know that I love you! Why would you set this up?" She screamed.

"The only thing I set up was for you and the Lieutenant to enlighten each other's shadows!" Jung defended.

"We could have it all! Companionship and passion; our minds will feed each other's intuitions! Together we can do great things and have a blissful life!" Sabina poured her heart out to Jung.

"Sabina, it is not that simple! I have a wife!"

"You said you love me more than you love her!"

"Don't tell me what I said!" Jung snapped. "That does not mean we can just run off into the sunset and all our problems will disappear!"

"What problems would we have?" Sabina challenged. "I know what we have is real, so tell me why you deny me."

"Do not act like a silly little girl with stars in her eyes! The world is not so simple, Sabina! There are financial concerns. How would we pay for this fantasy life you are dreaming of?" Jung said waving his hand in the air.

"We love each other so we'll find a way!" Sabina pleaded.

"If I am forced to worry about money then I would not be able to contemplate my theories as I do." Jung explained in vain.

For a few moments a painful silence filled the cluttered office. "Your wife is rich…" Sabina trailed off.

"My reputation would be tainted all across Switzerland, I would no longer be trusted as a professional in my field. Where would I go? What would I do for money?" Jung reasoned.

Sabina just shook her head and sobbed. "If I was rich then would you love me?"

The two looked up as the sound of the office door slammed shut interrupting them. Neither had noticed Rupert getting up and leaving.

"You naive girl!" Jung snarled angrily. "How dare you tell me what is happening between us! You think that you understand how the world works but you don't know the first thing about it!"

"I gave you everything! I left my home to study with you, I loved you; I still love you!" Sabina pleaded until she felt sick to her stomach.

"And you gained much!" Jung lashed out.

"Who are you? I feel like you are someone I've never met! You are supposed to be here for me, to help me through the hardest part of my life. I will never forgive you for this! Not in this life or the next!" Sabina spat before storming out of his office.

Two Months Later

It was a cool, cloudy afternoon as Rupert sat in the grass in front of St. Bees Lighthouse watching a pair of ospreys dive into the water. His father was active in the community and got him a job with the St. Bees Football Club. It seemed like a good opportunity, even for a man with one leg, but as he watched the waves gently bobbing up and down in the dull light of the overcast day, there were other things on his mind. He thought back to the last conversation he had with Sabina, the night before he left Switzerland.

"Will I ever see you again, Sabina?"

"I suppose that is not for us to know. My grandmother would have said, im yirtzeh Hashem. "

"That's beautiful. What does it mean?" Rupert asked.

"Something like, if God wants it to happen then it will." She explained.

"Ah yes, back in India they have a similar saying, bhagwan kare, I think it was." Rupert paused to reflect on everything. "I know I didn't kill my men, but I still find it hard to actually beleive it. I know I need to make peace with it but it's like there will always be a part of myself I cannot forgive. My nightmares have subsided, but the feelings linger."

"You have your whole life to forgive yourself." Sabina replied.

"And yet, somehow I know that at the end of my life I will still carry the guilt I have for letting William die. Have you forgiven yourself?"

"I'm afraid I have more things to address before I can forgive myself, but one day I may have a peaceful mind."

"We could have been good together, you and I."

"I know." Sabina said calmly. "I feel our destiny was never to be lovers. Maybe we are just supposed to be friends."

"Our destiny..."

Rupert's memory faded back to reality in his seaside village in Northern England.

Those words, and the way Sabina looked as she said them, were seared into Rupert's mind. He sat in the grass, staring out over the ocean and thinking about life.

I think it's time for me to stop worrying about what destiny has in store for me and start living the life I want for myself.

He reached in his bag and pulled out an old bible. The pages randomly fell open to Psalms 46, which he began to read. *God is our refuge and strength...*

After his experience, Rupert wasn't sure if he believed in God as other Christians did, but he knew that reading his Bible brought him comfort and that's all he needed to know.

He looked out to sea and saw an osprey come up from the water clutching a fish that wriggled helplessly. Death comes for us all, I suppose. And yet, here I sit, in one of my favorite places in the world. I have life, what shall I do with it?

For the rest of his life, Rupert studied Dr. Jung's work. Even though things turned out the way they did, Rupert always felt a sense of connection to the Doctor's ideas. He felt like he was still contributing to them, in some indirect way. More than once, he thought about writing to Dr. Jung, but never did. He became content remaining an archetype in Jung's memory. [FACTS 32, 33]

FACTS

32.   William Leefe Robinson went to the Western Front a second time, and again, was wounded in battle, being shot down during bloody April. For the rest of the war, he was imprisoned in multiple German POW camps where he was regularly punished, both for his notoriety as the man who shot down the first zeppelin, and for his failed escape attempts. He had four serious escape attempts which did not deliver him to Switzerland, or out of Germany at all. Instead, the only place he went was solitary confinement. This lasted until the end of World War I.

He never met Dr. Jung or Sabina. Upon returning home in December of 1918 he had all four limbs, still very much attached to him, but he was extremely weak from his imprisonment. He spent Christmas with family and friends and then contracted the Spanish Flu. Leefe died on December 31, 1918, at the age of 23. [Ref 24]

33.   Sabina Spielrein began as Dr. Jung's apprentice in 1905. They broke off their affair which lasted 1909, the exact nature of which is up for debate but we do know that it was putting Jung's marriage in jeopardy. In 1911 Sabina received her doctorate and published the first psychoanalytical dissertation by a woman.

In 1927 Sabina discovered her husband was living with another woman and even fathered a child with her. She took their two daughters and left him for her home in Rostov-On-Dom, Russia. In 1936 Stalin banned psychoanalytics from the Soviet Union along with many other academic fields of study. Communism stole her profession.

In 1942 Nazis invaded Sabina's hometown for the second time. This time they identified all the Jewish people, including Sabina and her two daughters. They were herded out of town and murdered. Fascism stole her life.

Sabina was a trailblazer for women in academia. She authored over 30 publications, many of which were original studies in the field of psychoanalytics. [Ref 25]

Leefe and Sabina deserved better. But in life, what we deserve is not what we get. Instead, we only get what we create, regardless of whether that creation is through our own actions and inactions or the actions and inactions of others. People can be ruined by atrocities they had no hand in creating. Likewise, people are capable of preventing atrocities by standing up against evil ideas long before they are acted upon. After all, it is our free will that designs our individual and collective realities.

Are you living the life true to the world you want for us?

Preface to
Chapters 14 through 34

Have you ever wondered what creates the universe and our experiences in it? What if you could manipulate the tiniest building blocks of reality? Would you influence them to make your life better? Perhaps you are more curious as to what is on the other side of the veil that separates the world of the living from whatever lies in the great beyond.

It might seem like time itself is steadily pushing everything in a certain direction and yet the great Albert Einstein said, "the distinction between past, present and future is only a stubbornly persistent illusion."

If you want to explore the deepest truths of our reality and learn how you can use them to your advantage, then illuminate your curiosity with the following chapters that will take you out of this world...literally. You may find concepts you are familiar with but you will also find things that push your understandings to the limits of the universe, including your place in it.

The Law of Attraction is real but is not an explanation of how the universe works. Rather it is a methodology you can use to create the life you want for yourself. Your consciousness is part of the universe's creation process. You are helping to create reality with your actions and thoughts because you physically exist as a part of it. If you're looking for an explanation of the universe and how the Law of Attraction could possibly be true, then read on and find the knowledge you seek.

The next few chapters will act as a general introduction to foundational concepts in chemistry, physics, and psychology. As you continue reading the ideas will challenge the prevailing theories with abstract possibilities of what might explain the scientific mysteries that remain unanswered. It is best to read these chapters when you are mentally attentive and prepared to listen-in on an increasingly complex conversation.

You are about to be introduced to the entirety of reality as we know it. From there, it is up to you to explore the mysteries on the other side of the veil that separates this life from the great beyond.

Chapter 14

The Lone Wolf

"I am not what happened to me,
I am what I choose to become."

-Dr. Carl G. Jung

After World War I, Dr. Jung rarely thought back to his conscripted service at the hospital. He felt his romantic connection was better left in the past, or perhaps, forgotten altogether. Instead, most of his memories from the field hospital were the wholesome, but otherwise insignificant, interactions with patients.

The memories of Rupert' experiences, however, were not so easily forgotten. Jung was grateful for the deep insights into consciousness he got from Rupert, but he did his best to forget about the personal interactions they were all tangled up in. Missing the opportunity to study a near death experience was quite regrettable. To Jung, Rupert became less of an individual and more of an enigmatic, larger-than-life figure who was somehow also reduced to a faceless, nameless patient from long ago.

In truth, thinking of Rupert illuminated the shadow Jung had created to cover his infidelity, so he blocked it out to maintain the persona of a good husband.

Dr. Jung moved home to Zurich and settled back into life with his wife, Elizabeth. He figured he'd get back to teaching at a university but instead reopened his clinical psychiatry practice which became the focus of his life. The practice provided Jung many opportunities to study the mysterious unconscious mind by helping clients shed light on their psyche's shadows. It was one of these patients who would help Jung conceptualize and mature some of his most controversial ideas.

14 Years Later, 1932

"Your new patient is here, Dr. Jung." His assistant said, leaning through the doorway. Jung acknowledged her with a gesture as he put his circular glasses back on his face, which was beginning to show his age. A stocky, disheveled man who looked as though he had been in a fight, briskly walked to Jung's burgundy lounge chair before plopping down dramatically.

"Good morning. You are Herr Wolfgang Pauli?" Jung said, looking up from his appointment schedule. The man smelled as though he was sweating alcohol. He looked relatively young, probably around 30 years old but his hairline had already receded, leaving his forehead in the shape of an 'M' stretching from one ear to the other. His hair was shortest on top and got longer as it moved away from the center of his head, tapering off towards his ears. This gave his whole head a rather square shape that was exacerbated by his dull expression. His eyes seemed to droop even though his eyebrows remained slightly raised. The man needed a shave, but Jung wondered how much of his current appearance was due to his drinking and how much was due to the reason he was drinking in the first place.

"Yes, I am Pauli." The man said with a rough and boisterous voice that projected his big personality through the room. "I hope you can help me get my life back in order." He said directly.

"I hope I can help. First, I'd like to ask; who are you? That is, how do you define yourself?" Jung began.

"I'm a divorced man and a failure of a son. My wife left me. We weren't even married a year." He started. "I used to confide in my mother, whom I was very close to, but she recently took her own life. I am overcome with guilt and I'm at my whit's end; I have nowhere left to turn." [FACT 34]

He is giving me a lot to work with, Jung thought optimistically. Now to help him figure out the true cause of his troubles; what is he needlessly blaming himself for and what has he blocked out that he should take responsibility for. Jung noted that Pauli seemed like a well-dressed man, or rather, he would be well dressed if he had a sober enough mind to tuck his shirt in and fix the pant leg that was bunched up, revealing an old sock.

"You have come to the right place. I think you have also turned to alcohol to help cope with all this. I will ask, as we continue your therapy, to cut back as you are able."

"Mm hm." Pauli huffed before pulling out a small comb to address his messy, jet-black hair.

"That can be a difficult task, so above all else, be honest with me. I am not here to judge you, and I will not speak to anyone about anything you tell me, you have my word as a doctor." Jung finished his introduction.

"Thank you." The man said with a sigh, "I can already tell we'll be friends."

Friends? Jung thought to himself, that must be the alcohol talking.

"Why don't you start by telling me about you and your mother's relationship when you were young." Jung said.

Pauli described bits of his childhood and spoke of some of his more recent troubles. The session came to a close but Pauli wanted to know his prognosis.

"How many weeks of therapy do I need?" He asked directly.

"When I counsel you, I keep two goals in mind. The first is to help you cope with your immediate grief while the other goal is to get you adjusted to your new life." Dr. Jung instructed.

"And what new life is that?" Pauli blurted out.

"Whichever one you define for yourself. Today, you walked in here defining yourself as a divorced man and a failure of a son, but throughout this counseling you will create a new personal identity for the next chapter of your life."

"I don't understand. I am not here to be reborn; I am here to cope and grieve."

"Sometimes coping can make you realize you are not the same person you once were." Jung said authoritatively, pausing for dramatic effect. "I cannot predict how many weeks or months you will need to see me. Grieving is unique for each person and sometimes when it seems you are through it, something triggers a relapse, and it feels like you are starting the whole grieving process again. You will always be someone who lost his mother to suicide but you will come to realize that you are not defined by the things that happened to you. Instead, you can choose exactly what defines you." Jung thought about common ways people define themselves and asked. "What is it you do for a living?"

"I am a physicist." Pauli said.

With great enthusiasm Jung said, "really?!" Immediately distracted from the point he was making. "That is a fascinating science! Can I ask; what is your take on these new quantum theories they are coming up with? I mean to say, I don't fully understand them, but I hear nothing makes sense anymore."

"Ha! That might be the most accurate description I've heard!" Pauli responded. "What do you want to know?"

Jung jumped straight to a burning question. "Can free will exist or is the universe deterministic? In other words, is everyone's entire life already set or can people influence their futures?"

"Destiny or free will; very philosophical of you. The answer is..."

Dr. Jung was on the edge of his seat while Pauli enjoyed the suspense. Finally, he continued.

"It depends who you ask. The folks in Copenhagen think true randomness exists." Jung gave a confused look so Pauli clarified, "which does allow for free will. There is a Frenchman who believes there are pilot waves which remove the randomness."

"Meaning everything that will ever be, has already been determined and there is nothing anyone can do to change it?" Jung clarified.

"That is the common understanding of destiny, isn't it?" Pauli said pretentiously. "In that theory the pilot waves would influence how the quantum particles localize."

Jung had no idea what that meant and Pauli didn't explain.

"Then there is the Many Worlds interpretation which allows for free will but requires an infinite amount of energy because anytime there is any uncertainty the entire universe splits in two so that both possibilities physically exist without any knowledge of each other." Pauli happily explained with a pep in his voice that was nowhere to be found during the therapy. [FACT 35]

Dr. Jung was keeping up the best he could. "Interpretation? Meaning there are different explanations of the same phenomenon?"

"The same phenomenon... yes," Pauli thought aloud, "I think that is a good word to describe it. Do you know which phenomenon I'm talking about?"

Jung was floored. He had no idea what Pauli was about to say. Clearly it would be something profound, some seminal part of our existence.

"The phenomenon being interpreted is nothing more than mathematics." Pauli said. "Math is the common language of our universe."

"Mathematics?" Jung was disappointed. "If math describes our universe, then does that mean math is discovered rather than invented?" Jung asked.

"Absolutely. Mathematic symbols are human inventions but what they symbolize is the most absolute way to describe our reality, regardless of human interpretation."

"And somehow this math can be interpreted in different ways?" Jung postulated.

"Well of course!" Pauli said jovially.

Jung was astounded at Pauli's new demeanor. Far from depressed, he seemed truly happy. It was as if the person sitting through therapy for the last hour had left and a new person was sitting before him. People often mask their grief, Jung thought.

"I thought math was absolute. How can people think it means different things." Jung thought aloud.

"It gets complicated when some formulas become dependent on other formulas which could have two possible solutions, then one assumption leads to another, and we are all tangled up in a web of theories which can go one way or the other." He explained. "What is it you want to know exactly?"

"I want to know it all." Jung said blankly. He didn't care about the math or the physics. Instead, Dr. Jung was on a quest to find the deepest secrets of the human experience in this wonderful thing we call life. He understood modern physics was starting to weigh-in on things like destiny, so he jumped at this opportunity.

"It sounds like you want to understand quantum mechanics." Pauli stated.

"Alright, what is quantum mechanics?" Jung asked as if it was some simple question like, what is the weather.

"Quantum describes the discretization of waves of energy."

Jung had no idea what Pauli just said.

"There smallest building blocks of reality have an inherent uncertainty." Pauli replied concisely.

Jung tried to keep pace saying, "You cannot define the smallest bits of reality, why is that so strange?"

"It isn't that we can't define them," Pauli said mystically, "it is that the particle itself is not one thing or the other; it is simultaneously both blue and red, up and down, here and there."

"That doesn't feel like science." Jung blurted out. "It feels fluid, as if the very foundation of reality could change at a moment's notice, controlled by nothing more than a feeling. I thought science was rigid." He added before Pauli dove into a lecture.

Pauli didn't mind talking about his work, in fact it felt more therapeutic than the therapy itself. "What is the most basic way you can describe your profession?"

"Hm," Jung thought, "I use an analytical approach to study the human experience by focusing on what causes individuals to have certain feelings and do specific actions." Jung said. "I then compile data from many individuals to create theories for large groups."

"No, no, no!" Pauli blurted out as if he was scolding a student. "That's too complicated. Reduce it, make it simpler." He demanded.

"Oh. Well, I study why people feel and do the things they do." Jung said feeling very dissatisfied with this abbreviated answer.

"Good! Now let's ask the physicist the same question. What is physics?" He asked.

Jung thought about the possible answers which all seemed quite complicated.

"Physics describes reality." Pauli stared directly at Jung in silence to let it sink in. "That's it! The physicist has two ways to describe our universe, mathematics and theories. The theories, of course, are based in mathematics." Pauli explained.

Jung nodded.

"Before Isaac Newton, they thought the laws of physics on Earth were totally separate from the laws of physics in outer space. When the apple fell, Newton thought up this thing called gravity which simultaneously explained the physics on Earth and the physics that pulls the planets. His theory unified all of physics with one set of rules, true in all cases. For the past few hundred years, that was that, but now that we are dealing with particles smaller than atoms, things are different." Pauli said.

"The smallest bits of the universe don't follow the same rules as everything else?" Jung confirmed.

Pauli nodded, affirming, "We now require two sets of laws to explain our reality, just as we did before Newton. You can see the pattern."

"What pattern?" Jung said. "Are you saying the current science is wrong? You think someone will come up with a new theory to unify physics again?"

"In short, yes, but of course, it's all very complicated."

"How so?"

"You are a very curious mind-doctor." Pauli said, intrigued that his therapist seemed genuinely interested in physics. "We don't know what we don't know. It might take a hundred years just to define the questions. Then we may be able to hypothesize the answers, but yes, I believe there is only one set of laws and whoever can prove it will be just as famous as the great Isaac Newton."

"That's frustrating." Jung said without thinking.

"Ha! I would agree. Why are you so interested in physics Dr. Jung?"

"I have always wanted to understand everything there is to know about the human experience. In all my contemplations I seem to get stuck on two core questions." Jung explained.

"And what questions are those?" Pauli asked curiously.

"Where did the universe come from and what are the limits of consciousness."

"Why would consciousness be anything more than what we experience?" Pauli challenged. [FACT 36]

Jung chuckled. "Throughout your therapy with me you will see firsthand how your own conscious experience is just one part of your mind. You have a vast unconscious mind that connects each of us to the world in a very real way."

Pauli was skeptical and became defensive of his personal issues, so he brushed it off and steered the conversation back to physics. "Well, I cannot tell you where the universe came from, but I can tell you where its coming from." He said with a sly grin.

Jung looked puzzled, not sure if Pauli was playing word games or if there was an actual difference.

"Before I get into all that," Pauli began, "you should understand reality from different perspectives. There is one reality, but it looks different at different scales. The most important perspective to us is what I'll call human-reality. We exist at this scale of reality; it is everything you know."

"Our consciousness must be part of this human-reality." Jung claimed but Pauli spoke on as if he hadn't heard him. [FACT 37]

"Much larger than our humble perspective is the cosmic scale, and as you just learned, the physics are the same even though the scale is vastly different."

"Cosmic meaning stars and planets and whatnot," Jung said. "Excuse me, but how many scales are there?"

"How many do you want there to be? What I called the cosmic scale could be divided into many scales: solar systems, galaxies, galaxy groups, superclusters, the intergalactic medium, the whole way up to the cosmic web. For this conversation, let's focus on four scales: cosmic (large scale structures), human-reality (macro), microscopic (atomic based), and quantum (sub-atomic)."

"I get it. There are more granular scales but for now I can just conceptualize reality as things bigger than us, us, the things that make us up, and the things that make up those things." Jung understood.

"We study large scales and tiny scales to figure out how the universe works but what is most important is using that knowledge to benefit ourselves at the scale of human-reality," Pauli explained. "Who cares if you can manipulate some tiny, microscopic organism?"

Jung was confused.

"If that microscopic thing was a germ that causes a human pandemic, then you'd care. Why should we try to understand how some star thousands of light-years away is imploding? If our sun was at risk of the same fate, then we would be desperate for knowledge on that subject."

"I see. You aren't saying other scales don't matter. You're saying the reason we delve into them is to manipulate things that affect us." Jung clarified.

"We are not the center of the universe, but perspective is everything. This entire lecture is framed around the perspective of human reality because we don't exist to benefit the center of the universe. Everything we do is for humanity."

"As a psychologist, I appreciate that." Jung said. "In a way, each of our conscious minds' are framed by this perspective of human reality. What I mean is, there is only one true objective reality."

"The reality everyone agrees is real." Pauli agreed.

"Yes, and since we all exist in this one reality, our individual consciousnesses are based on the physics you are describing. Our professions appear very different, but this is exactly why I believe they are fundamentally connected."

"Indeed," Pauli said. "As I continue this lecture you will find that not only your profession, but all professions, are rooted in physics because physics describes the conditions that allow for any existence at all."

FACTS

34.   In 1932, the future Nobel Prize winner, physicist Dr. Wolfgang Pauli was in a very tumultuous period of life as he had been married less than one year and his mother committed suicide around the same time, which left him living in mental anguish. Though he chose to divorce his wife, his life spiraled into crisis jeopardizing his career. Pauli found local psychiatrist Dr. Carl Jung, seeking psychiatric therapy. [Ref 28]

35.   There are various interpretations within Physics that lead to very different ideas of how our universe works. The three interpretations listed in this chapter (Copenhagen, de Broglie-Bohm, and Many Worlds) will be explained in future chapters after some other prerequisite explanations. (Particle-Waveform duality, waveform collapse/localization)

36.   Consciousness has been defined many times in many different ways. Even without a definition we all agree on, there is a widespread belief that we all intuitively understand what consciousness is because we all experience it. [Ref 29]
        The most honest and accurate definition was published in 2004 by a team of 8 neuroscientists. They admitted there is no scientific foothold for any real definition of consciousness because of its inherently subjective nature. There are no physical constructs of consciousness, no mass or even any massless particles that conjure it. [Ref 30]

37.   Human-Reality is not a term you will find in your textbooks. It is defined as our shared objective reality, at the scale we experience, acknowledging the human experience that contributes to it. In future chapters this concept will mature into an understanding of an internal and an external reality; referred to throughout this book as human-reality.

Chapter 15

How to get to Quantum

"First," Pauli lectured, still sitting in Jung's chase lounge chair moments later, "let us divide everything there is into two groups; there is the stuff in the universe, or *matter*, and there are *forces* which act upon that matter. We will begin by discussing the stuff from the largest scale to the smallest and in doing so, you will arrive at the doorstep of our quantum reality.

"At the largest, cosmic scale, we have things like stars, white dwarfs, and whatnot but most of the things out there are all separated by unimaginably large amounts of empty space. We will address this 'empty' space later. First let's talk about the things within space. Planets, asteroids, and other debris are all prevalent throughout the universe." Pauli explained.

"What about black holes? Those are giant things out there in space, correct?" Jung said.

"Actually, black holes may fall into the category of a force, we don't know for certain but either way, that is a more complicated discussion. Right now, we are talking only about matter." [FACT 38]

Jung's mind created an image of the universe at the cosmic scale. He was floating in outer space looking at a vast darkness with twinkling lights scattered in every direction, all of which seemed impossibly far away. For a moment Jung lost himself in the wondrous magic of the grand universe as if he himself was some aqueous transmission. As Pauli spoke, Jung's vision played like a grand performance he got to sit back and enjoy.

"Somewhere in all that space and stuff is the Milky Way galaxy." Lectured Pauli.

A spiraling river of stars and cosmic dust zoomed up from somewhere far away for Jung to admire. It was the Milky Way Galaxy, and it was beautiful. The magnificently spiraled disc of twinkling lights wisped into the darkness of space like fairy dust thrown into the night sky.

"If you zoom in on one edge of our galaxy you will find our solar system."

Jung's vision kept pace with the lecture and the celestial image magnified. At first, Jung noticed the whole river of light was made up of individual stars like one of those paintings where the artist uses thousands of dots to create an image. The magnification sped up and each speck of light grew larger, shooting out around him before spreading out so far, he could barely see them anymore. A giant red sphere flew by him and then the image froze. The blue and green Earth was gently rotating in front of him like a decorative glass marble hanging from an invisible string. He marveled for only a moment before Pauli continued.

"On Earth, we find our human reality, where we *experience* physics."

The Earth that Jung was marveling at grew to an immense size in front of him. He stood still as the atmosphere exploded out past him like a shockwave form an explosion. He flew right through some puffy clouds before seeing the familiar shape of Europe outlined by the Mediterranean Sea. His perspective continued zooming in until he recognized Zurich, then his own office building, and finally he was looking into the window of the room he was currently sitting in.

"Of course, you are already familiar with this macro world, as you are living in your own human reality right now. Let's continue to the microscopic world. Each of us is made up of cells."

For just a moment, Jung's imagination lingered in the very room he was sitting in and he became aware of himself for the first time since this vision began. It was like an artificial out-of-body experience where he could control his body but he wasn't in it. In an instant he was diving back into his imagination once more. Quickly, a spot on the back of his own hand came into focus. He saw what skin cells might look like. They resembled individual blocks, squished together to fit whatever shape the space would allow for. Each cell had the same components as the neighboring one.

"Cells are made up of proteins, fats, carbohydrates and other chemicals. These chemicals are all various arrangements of molecules."

Jung's imagination landed him inside a single cell. He saw an organelle rhythmically rising and contracting as if it was breathing. The cell felt slimy and wriggled like a living thing. It *is* a living thing, Jung thought to himself.

"Molecules are a specific shape based on what each molecule is."

"So, a saturated fat molecule is the same shape as other saturated fat molecules." Jung understood.

"The human body digests food by breaking down clumps of molecules. Cells then use the molecules in various ways."

Jung looked to the membrane of the cell wall which magnified until he could see a single molecule. It was a colorless buzzing of energy that mildly resembled a large sphere with two smaller spheres attached on one end, all harmoniously vibrating a low hum. All three were inseparable, one molecule, indivisible by human digestion.

"A molecule is made from atoms, bound together. Atoms are the smallest an *element* can be divided before it will no longer be that element."

The vision zoomed in closer to one of the buzzing spheres. The whole shape of three connected spheres is the molecule, he thought, this molecule has three parts, those must be the atoms. But what is an atom?

Before his eyes the molecule he was watching disappeared and all he saw was a periodic table, however, it was not complete. This whole vision was taking place in Jung's mind and he could not remember the specific data on the periodic table. Nevertheless, the chart had the distinct bowl shape and listed every element known to man. [FACT 39]

Jung studied the periodic table in his mind and thought, each element listed here is made of atoms, each element is the same type of atom.

"An atom cannot be chemically broken down; it is the primary constituent of human-reality." Pauli explained.

"That's why you described this scale as atomic based!" Jung had a eureka moment as he realized what Pauli was saying. "Everything is made of atoms and every element can be reduced to a single atom!" Jung blurted out, thinking about the spheres of buzzing energy, for the first time realizing what the things on the periodic table actually look like. "Molecules can be broken down to different types of atoms, but an element can only be reduced to one atom."

"Exactly. Take water for example. Its chemical composition is H_2O, which is a substance. Water can be reduced to one molecule, but it does not exist on the periodic table." Pauli explained.

"Because only atoms exist on the periodic table and water is more than one type of atom." Jung understood.

"If we break down one water molecule the water is gone, leaving its elementary parts; two hydrogen atoms and one oxygen atom."

"Single atoms cannot be reduced further, correct? That is why they are elementary." Jung confirmed.

"Close. Atoms *can* be broken down further but the stuff you will find inside the atom is no longer an element." Pauli said.

"Just like the water molecule. We broke it down into its parts and it was stuff totally different than water."

"Right but atoms are different. When you break down an atom you enter the *sub*-atomic world and once you go subatomic the periodic table is useless because no elements exist there."

"Subatomic meaning smaller than an atom." Jung thought aloud. "Smaller than any element."

"That's right." Pauli said. "It is at this point where classical physics begins to cease. From the atom up, we have one set of laws that accurately describe reality. Once we dive into an atom, we need a whole new set of laws to explain what is happening."

"Just like before Newton described gravity." Jung added.

"Let's go subatomic!" Pauli said, drawing a crude diagram of an atom. "All atoms are made of three types of subatomic particles; protons, neutrons, and electrons; like this."

Jung was still wrapping his head around everything they had talked about and now was forced to push all that to the side to keep up. Okay, he thought to himself. We are going subatomic which means we can throw out the periodic table. This is like a whole different reality from ours, Jung reasoned, in this reality the largest scale structures are atoms. All atoms are made of three types of subatomic particles, he repeated to himself.

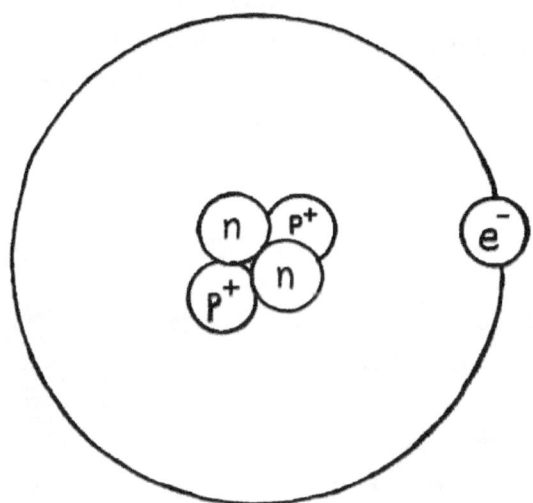

Pauli described his diagram as he drew, starting from the center. "Each atom has a nucleus at its core which is where the protons and neutrons are. Electrons are different, they orbit the nucleus."

"Just like planets around the sun!" Jung said.

"No." Pauli said shaking his head. "If you were to actually see these electrons, each one would look like a cloud rather than a single particle because they are constantly popping in and out of existence all around the nucleus." Pauli explained.

"What do you mean they pop in and out of existence?!"

"Ah, I've piqued your interest. Let me finish describing these concepts then we will get into that." Pauli said.

"Okay...please allow me to recap." Jung said, trying to think of a way to get through the lecture quicker. "Everything we experience in human reality is built up from atoms, in other words, our reality is atomic based. The periodic table lists every type of atom and the laws of physics are the same from atoms to celestial structures."

"To be clear, protons and neutrons also follow Newtonian physics, it is the electrons that follow quantum rules." Pauli interjected.

"Okay," Jung continued, "Every atom, regardless of whether it's gold, or hydrogen or whatever, is made of these three types of particles." Jung said as the realization set in. "So, you're saying that every different element is made up of only protons, neutrons, and electrons? Gold and silver are made of the same thing, and so is oxygen and... everything else?"

"Only at the subatomic level, but yes, all the matter, the stuff, of the entire cosmos and even our own bodies are made up of different atoms which are only protons, neutrons, and electrons."

"How do these three magical particles become different elements?" Jung asked.

"It all depends on how many protons each atom has. Sodium has 11 protons and 12 neutrons in its nucleus. It then has 11 electrons orbiting it. Another atom with the same number of neutrons and electrons, but only one more proton, is magnesium."

"Okay." Jung said as if he was waiting for more information.

"Do you know why the periodic table is arranged the way it is?" Pauli asked.

It's more than a random shape, Jung thought, I should have guessed. "No."

"All the elements are arranged by their atomic weight."

"What does that mean?"

"How many protons each atom has!" Pauli explained as he began to label his drawing. "That's what the little number in the corner is. Hydrogen, one proton. Helium, two protons. Lithium, three protons. Beryllium, four and so on."

"I get it! The periodic table starts with the element that has one proton then just counts up. That means the periodic table's shape is actually a mathematical grid." Jung exclaimed.

Dear Reader,

The following section will introduce basic chemistry, provide a deeper understanding of the nature of electrons, and give an example of how atoms form molecules. If you understand that, feel free to skip to page 167.

"Very good." Pauli encouraged as he drew another diagram. "Now allow me to explain how electrons fit into all this."

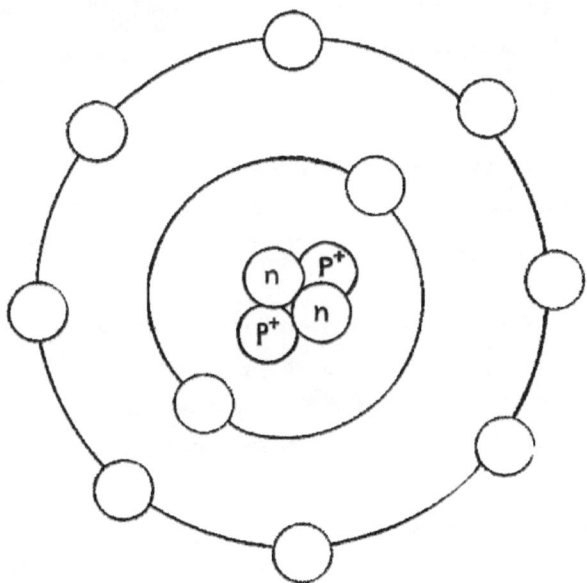

"What do electrons actually look like? Again, I can't help but see the resemblance to our solar system."

"These rings are just a representation. If you could watch an atom up close in real time, you would not see any resemblance to a solar system, it would look like a buzzing cloud of energy. The nucleus at the center would be totally obscured by the electrons in the rings."

"How?"

"In the solar system," Pauli explained, "planets orbit the sun in ellipses, almost circles, and they follow a trajectory based on classic physics along a single plane. In an atom the electron's orbitals are not just a single circle but a whole sphere. Actually, orbitals are 3D shapes that look nothing like a sphere. Finally, the way electrons move, when in an orbital, is unlike anything you've ever seen because they follow the rules of quantum physics."

"If its orbit isn't a sphere shape then what would it be?" Jung tried to understand.

"The orbital rings themselves are different 3D shapes based on which layers you are looking at. One could be like a ball that has been squeezed and then smashed on some sides while another is a dumbbell shape. There are a number of other shapes, but the shape doesn't matter for what we are talking about here. What I drew does look like the trajectory of a planet but it is just a representation." Pauli reiterated.

"I get it. This drawing is a two-dimensional representation of a three-dimensional atom that is paused like a photograph." Jung said.

"No, you're not getting it," Pauli said bluntly. "It is not so simple. Gravity causes planets to follow a predictable trajectory in a flat circle."

"You can guess where they will be in the future because of where they were in the past." Jung said.

"Exactly! When I said electrons move in a way you've never imagined, I was not talking about the shape of the orbit, I'm talking about the way an electron moves within that orbital. I don't want to confuse you but the odd quantum rules require the electron to be in all possible places in orbit simultaneously."

"Electrons can teleport?"

"No." Pauli laughed as if he enjoyed Jung's frustration. "Each orbital ring as a whole is simply hosting the energy from the electrons. In other words, a single electron particle isn't really a particle at all."

"Now I am thoroughly confused, yet quite intrigued." Jung said.

"I'd be surprised if you weren't confused!" Pauli exclaimed.

"First you tell me particles can pop in and out of existence, now you are saying this particle isn't a particle at all. This must be the quantum weirdness I've been waiting for." Jung said.

"These drawings are just representations. The important part of that drawing is the math associated with each electron-ring."

"But I don't see any math in your diagram."

"Perhaps you didn't notice it because it is so simple. Look again and count the electrons."

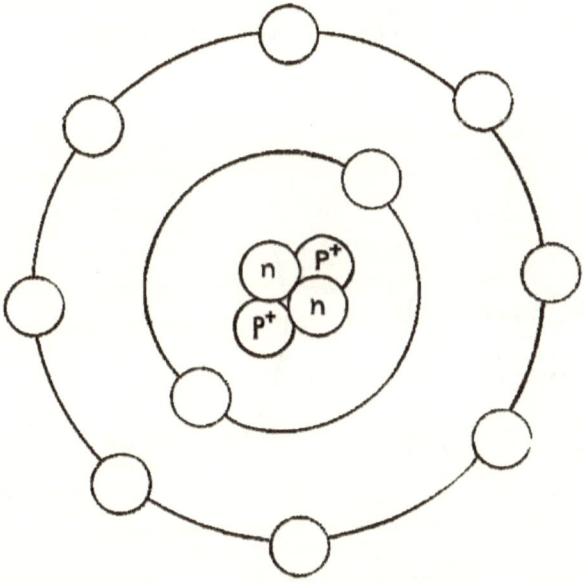

"Just count? On the surface this seems like one of the most complicated subjects in science but you are making it as easy as one, two, three." Jung said jovially.

"This is basic chemistry. Each orbital ring has a maximum amount of energy it can hold. The first orbital ring can support the amount of energy that two electrons have. If you have an atom with one electron and you add another, the second electron combines its energy with the first throughout the entire orbital."

"Like a forcefield? You described an electron as a single particle but now you are saying it's just energy that combines with other energy; is it a particle or is it energy?" Jung asked.

"Excellent question Dr. Jung! In a very frustrating way; it's both. You are jumping from chemistry to quantum physics. At this point, all you need to understand is there is a maximum number of electrons each ring can hold. The first ring's limit is two, so if we add a third electron, it naturally exists in a second orbital ring, farther from the nucleus in the center. The second and third orbital rings can each hold up to eight electrons."

Jung suspended his confusion about the duality of the energy-particles so he could focus on what Pauli was telling him. He looked back at the drawing to follow along with the lecture.

"Here is an atom with two electrons, and here's one with three." Pauli said.

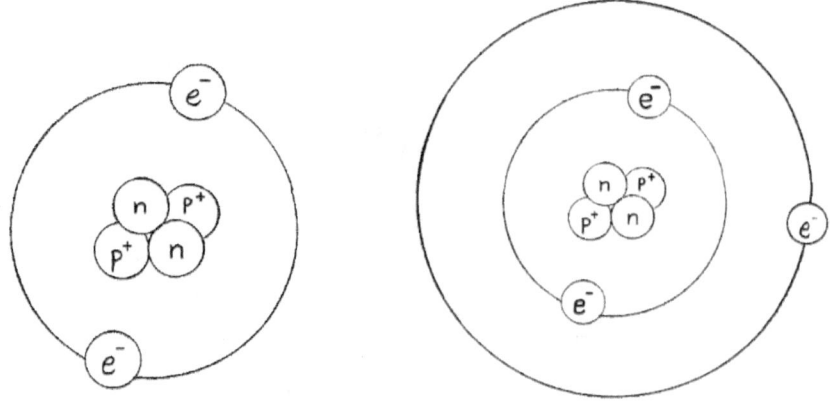

"Sodium (Na) has one lone electron in its third ring." Pauli explained. "Now look at chloride (Cl). Chloride has 17 electrons in three rings. Two in the first, eight in the second, and seven in the third ring. How many more electrons could fit on chloride's third ring?" Pauli seemed to be in his happy place.

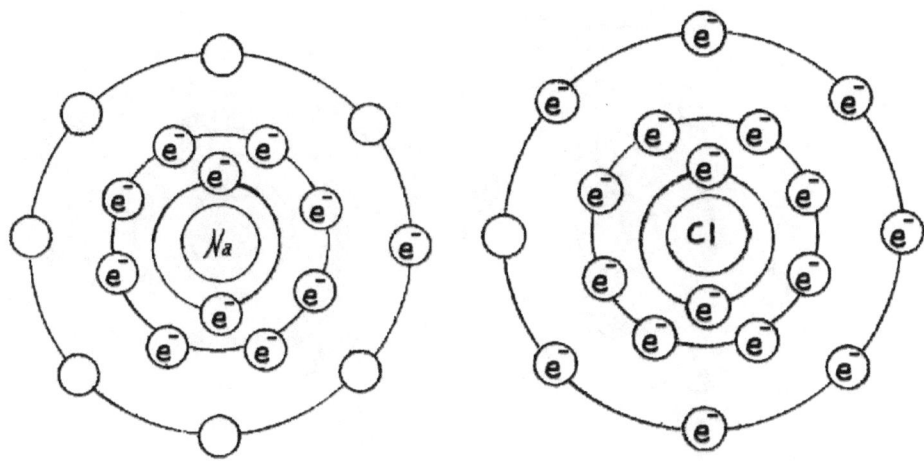

"One." Jung said, looking at the hole in the outermost orbital.

"Correct. Chloride's outermost ring has one opening left and sodium's outermost ring has only one electron present. You should not be surprised that these two atoms naturally bind together. The lone electron is pulled into chloride's third ring, filling it."

"When the electron binds to the chloride, is it not lost from the sodium?" Jung asked.

"Not at all. Remember adding an electron combines the energy to the energy already on that orbital. Think of the orbital like a force field that surrounds the nucleus. The more electrons on that orbital, the stronger the force field is. When two atoms combine in this way, the outermost orbital becomes a single force field surrounding both nuclei."

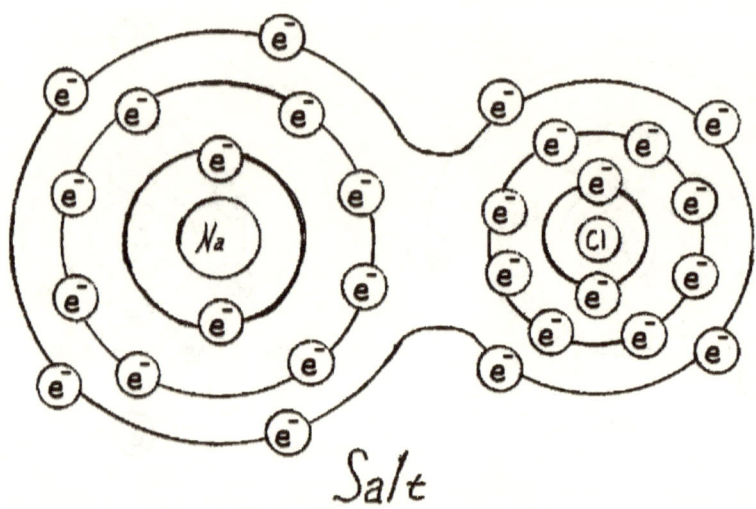

Salt

"Which creates one molecule!" Jung exclaimed. "Are all molecules formed so perfectly?" Jung asked.

"No."

"Then it is no coincidence that salt is necessary for our bodies to survive." Jung reasoned. "I imagine these two atoms began forming these connections, perhaps, hundreds of millions of years ago and over the course of our evolution, salt was abundant, and our bodies evolved to benefit from it."

"I like the way you think. This type of bond began happening billions of years ago. I'm no expert in evolution so I cannot say, but it would make sense that our bodies evolved to benefit from something naturally prevalent in the universe." Pauli figured.

"Let me see if I understand so far." Jung said. "The very smallest bits of everything that exist are just these three particles: protons, neutrons, and electrons. These three particles bind together to create atoms. Atoms come together to create molecules and molecules build upon each other to form everything we interact with, including our own bodies and brains."

"All correct. You are a quick study, Doctor Jung." Pauli said.

Jung started feeling confident he had at least a basic understanding so he said, "if I have four protons, four neutrons and one electron that could be one kind of atom, like hydrogen. Whereas four protons, three neutrons and two electrons would be another type of atom, like oxygen or something?"

How to Get to Quantum

"You are overthinking it." Said Pauli, glossing over the incorrect number of particles for hydrogen and oxygen. "Remember, it is only the amount of protons that effect which element the atom is expressed as. The number of neutrons and electrons effect other aspects of the atom, like how that atom interacts with other atoms. There is a much deeper explanation about isotopes and ions but all that is beyond the scope of this discussion."

Dear Reader,
If you skipped ahead, pickup from here.

"So, every different element is made of the same three particles and the only determining factor in which element the atom becomes is the positively charged particles? Interesting." Jung said as if he was lost in a thought.

"Precisely! Of course, there are many details I skipped but the entire field of chemistry explores the world at the atomic and molecular levels. It would be far too much to cover all of chemistry and physics." Pauli said.

"You explained that molecules are atoms combined by sharing electrons, but how do protons and electrons combine to form a single atom?" Jung asked.

"You are ready to go quantum!" Pauli said gleefully. "There is more to it but in a nutshell, there are fundamental particles that make up protons and neutrons."

"And electrons?" Jung asked.

"Ah, nothing escapes you! Electrons *are* fundamental particles." Pauli explained.

"Oh." Said Jung, trying to understand why electrons are fundamental, but protons and neutrons are not.

"Fundamental particles are the absolute smallest bits of matter in the universe." Pauli explained.

Jung remembered how this conversation began in the first place. Pauli broke everything into stuff and forces, and they were still just discussing the stuff, but it sounded as if they were reaching the smallest bits.

"So, protons and neutrons are made up of some other fundamental particles." Jung started.

"Quarks." Pauli said.

Jung stopped speaking and looked quite puzzled as he tried to understand what Pauli just said. "Corks?" He asked, worried his new client was asking to uncork a drink because he was having withdrawals.

"Ha! Not corks, quarks." Pauli said, enunciating the *kw* sound the *q-u* make. "Up-quarks and down-quarks are the fundamental particles that form protons and neutrons."

"If they are made of the same things then what is the difference between protons and neutrons?" asked Jung.

"All protons are two up quarks and one down quark while all neutrons are two down quarks and one up quark," Pauli explained.

"Okay, up quarks, down quarks, and electrons are the fundamental particles but if that is the smallest building block of reality then where do they come from, are they always in existence somehow?" Jung asked.

"I'm impressed Doctor Jung! First of all, there are other fundamental particles, but we will address those later. These quantum particles are the smallest building blocks… *of matter.*" Pauli paused for dramatic effect. "At this point, we can stop talking about the stuff of the universe and switch to forces."

"So instead of asking where fundamental particles come from, I should ask; why do these quantum quarks bind together to form protons and neutrons?" Jung figured.

Thunder cracked outside. A storm had rolled in but the men were too deep in conversation to notice. The boom of thunder had muffled the sound of Jung's assistant knocking on his door.

"Doctor Jung, your next client is here." His assistant said, startling the men.

"I suppose this will have to wait until my next visit, Doctor."

"I am looking forward to it!" Jung replied.

[FACT 40]

FACTS

38.   The idea of a mass so dense light could not escape its gravitational pull was first postulated in 1783 by a man named John Michell. At the time the idea was little noticed and had no real application.

168

The idea gained application in 1916 as a result of interpreting Einstein's theory of general relativity, however, they were called frozen stars. Wheeler first used the name 'black holes' in 1967 when he revolutionized the concept of what this dense mass might be.

This is the perfect example of when the math has two solutions. The math that insinuates black holes also creates the possibility for white holes, which would be a cosmic structure where matter emerges from. There are no observed structures like this, so this part of the math remains disregarded.

To claim that a black hole is a force is a controversial statement today and probably incorrect. Related to a greater debate on the whole truth of gravity, some modern interpretations do suggest that black holes might be products of forces rather than matter. [Ref 31]

39.   The first concept of a periodic table was in 1869 by a Russian chemist named Dmitri Mendeleev. The periodic table matured throughout the 20th century until it included all known elements. Meaning, at the scale of human reality, the entire universe can be reduced to what is listed there, without adjusting any laws of physics.

In contrast, when we look to the things that make up atoms, we can only explain what is happening with an entirely new set of laws, quantum laws. [Ref 32]

40.   The chemistry discussed in this narrative was scientifically accurate. The field of chemistry is not just the study of matter (specifically individual atoms and molecules) but the study of *interactions* of atoms, molecules and, well, chemicals. Chapter 15 focused more on the structures of matter and gave examples of what binds the structures together rather than chemical interactions.

For anyone interested in chemistry you can expect to start by using simple algebra to define properties of individual atoms. The next thing to study would be simple interactions of atoms. These interactions are all based on three things: protons, neutrons and electrons. Each single elemental atom has one standard configuration, for example, magnesium. Normally it has 12 protons, 12 neutrons, and 12 electrons. If it has any more or less neutrons then it becomes a magnesium isotope which effects the mass of the atom. If magnesium has more or less electrons it becomes a magnesium ion which effects how that individual atom interacts with other atoms.

In basic chemistry, you will calculate the specific interactions of various atoms, molecules, and solutions. You would not be making any new discoveries at this level but instead you would be practicing chemistry as if you were learning a new language. Anyone can splash two solutions together and create a reaction, but only someone who knows the language of chemistry can make new discoveries.

Chemistry has many applications in the field of medicine. After all, your body is a complex group of organs which act and react to the presence and absence of different chemicals. Every medicine you could possibly take is a chemical. Whether you are drinking a cocktail of water, honey, and whiskey or taking a pill from a pharmacist; it is chemistry at work. [Ref 33]

Chapter 16

The Four Fundamental Forces

"The best that most of us can hope to achieve in physics is
simply to misunderstand at a deeper level."

-Dr. Wolfgang Pauli,
PhD in Theoretical Physics

The afternoon sun shone brightly through the windows of Jung's
third floor office which was nestled in a quaint neighborhood in Zurich. All
the buildings on that street were four or five stories with steeply slanted roofs
that resembled the jutting peaks of the Swiss Alps to the South.

The area was filled with the eclectic charm of the Victorian age.
Some of the buildings were old half-timber style, whose large wooden beams
were exposed in the walls, forming beautiful designs in the building's façade,
which was typical in Switzerland and South Germany. Other buildings were
brick or had smooth finishes. There were iron signs hanging over centuries-
old bakeries and murals of mountain pastures hand painted on large walls.
The corners of each roof were home to statuettes of lions or coats-of-arms.
Even the doors and shutters were intricately carved and decorated.

Again, Wolfgang Pauli sat upon Dr. Jung's burgundy lounge chair for
this week's therapy. When he arrived, Jung noted he was either sober or did a
much better job of hiding his drinking. However, as Pauli walked in, the

smell of tobacco smoke exuded from Pauli's neatly tucked clothes until the smell of stale cigarettes filled the office. The men had a deep discussion about Pauli's relationship with his mother and how he met his wife, until the bell rang out signifying the end of their one-hour session.

"That is our time for today, I think you have made progress." Dr. Jung said optimistically. "How do you feel?"

"Yes, I might be able to see the light at the end of the tunnel." Pauli replied with a sigh. He looked to the door then back to Jung before he authoritatively said, "Now where were we with your science lessons?"

"I was hoping you'd bring that up." Jung said sounding relieved. "It would be improper for me to force that upon you."

"Nonsense," said Pauli, "unlike my students who sit through lectures, you have your own knowledge to offer in return. Now, where were we?"

"We discussed everything from the cosmos down to the particles that make atoms, but I believe you completed your lecture on the physical stuff." Jung offered.

"Yes, very good, very good." Pauli said as he gathered his thoughts.

"I believe you finished with the elementary particles that make up the protons and neutrons." Jung said confidently.

"The *fundamental* particles, you mean. Elementary is atomic and fundamental is quantum." Pauli said in a matter-of-fact voice that commanded the room. "I taught you about all the matter and now we are ready to talk about the forces which act upon that matter." Just as before, Pauli's dismal therapy-voice had abruptly changed to a voice with great confidence. He found a blank piece of paper and began drawing. "There are only four fundamental forces we know of – gravity, the strong nuclear force, the weak nuclear force, and electromagnetism."

"I'll start with an easy one." Pauli said.

"Easy; you must be starting with gravity, that's the only one of the four I am familiar with." Jung thought aloud.

"Actually, I will save gravity for last." Pauli explained. "Gravity is mysterious and might be much more complicated than it seems. We will start with electromagnetism or EM for brevity. I already hinted at it last week when we were discussing what holds multiple atoms together, do you remember what I told you?"

Multiple atoms meaning one molecule, Jung thought about the previous week's lecture. He recalled the diagrams Pauli drew when he was explaining how atoms combine. In his mind's eye Jung saw two different atoms sharing their outermost electrons like a force field which created the salt molecule. "They share electrons." He recalled.

"Exactly, but let's take it a step further. *How* do they share electrons? For that matter, how does one atom by itself have an electron at all? The answer is the electro-magnetic force."

"Perhaps I have heard of this force." Jung said as he searched his memories. "When I was a child, Edison and Tesla had a greatly publicized feud about it; electric power, correct?"

"No." Pauli said flatly with a pause. "Well, partially, but interestingly enough the electric current you speak of actually creates a magnetic field."

"A magnetic *field*? What do you mean by field?" Said Jung, rather intrigued.

"Excellent question, but we mustn't skip ahead."

"Right; how do atoms bind together?" Jung said, staying on track.

"Electrons always have a negative charge and protons always have a positive charge. Think of a magnet." Pauli explained.

"Opposites attract!" Jung said.

"Now you're getting it. Electrons and protons have electric charges which inherently cause them to interact with the magnetic force which pulls them together."

"Okay. I am starting to see what's going on here. Electro-magnetism is electricity and magnetism together and in each atom there are positive and negative particles that attract like a magnet, very simple indeed." Jung said.

"Simple so far, but magnetism is only half of the EM force. Now to your comment on Edison and Tesla. The electricity you are thinking of; how does it get from the source to the destination?"

"Some sort of wires?" Jung thought aloud, "yes, cables."

"Correct!" Pauli drove on quickly. "The wire might be made up of copper atoms. Copper, and other transitive metals, have one electron in their outermost orbital ring. Once we add some electricity that outermost electron is pushed down the line being transferred from one copper atom to the next."

"Giving us the flow of electricity." Jung understood. "But how is that related to magnetism?"

"When electrons flow in a current, a magnetic field is generated outside the wires. It is not possible to separate electricity from magnetism fundamentally because electrons are negatively charged, so we have the electro-magnetic force."

"Intriguing," Jung said, "that seemed pretty straightforward."

"Here is where I will confuse you. Electromagnetic energy is on a spectrum that propagates many things you are familiar with. Have you heard of the radio?"

"Of course." Jung stated confidently. "Wireless communication over long distances."

"Radio waves are pure electromagnetic energy," Pauli explained. "EM energy is also what transmits all visible light, microwaves, x-rays and others. They are all just different frequencies, or energy levels, of the EM force."

"When you say it transmits, what do you mean? EM energy carries visible light?" Jung asked.

"Visible light is pure energy. The EM force is pure energy. What I'm saying is that visible light *is* the electromagnetic force."

"Okay, this is not so simple after all. You are telling me that radio waves I can listen to are the same thing as light I can see? Which is also, somehow, magnetic electricity?" Jung said, almost alarmed at how his perception of reality seemed to be crumbling away before his eyes and ears.

"Before you hurt your head let me point out that *you* do not hear radio waves. Your radio receives radio waves and uses electricity to change them into the sound waves you do hear."

"Right." Said Jung, feeling a slight relief from the onslaught of new information.

"Let me draw another diagram."

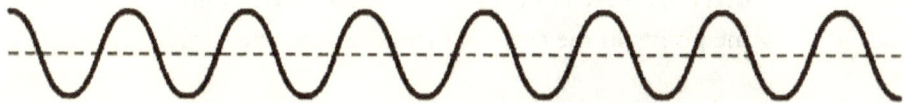

"This one is a basic wave. You can think of it like a wave on the ocean. If you are at the beach, standing in waist-high water and a wave hits you, you feel it, right?"

Jung imagined himself at the beach. He was standing in the water and watched as the top of a wave hit his body, pushing him back towards the shore. "I would get pushed backwards, yes." Jung said.

"Well in that scenario, what is the wave?" Pauli asked without providing further context. Jung thought about what Pauli meant but wasn't sure.

"The wave is made of water." He said.

"No," Pauli said, "the ocean is made of water. The wave is energy the ocean possesses."

"Oh, I've never thought about it like that."

"When the wave hits you, its energy is being transferred from the ocean to your body. Likewise, in physics, when we draw a wave like this, it is a representation of energy." Pauli explained.

"A visual representation of a basic mathematical description. It's like a picture of math." Jung thought aloud.

"Now a challenge; on the picture I just drew, please indicate one complete wave." Pauli instructed.

Jung took a moment to think about the wave of the ocean that pushed him backwards. The water got higher as the energy was carried and the top of the wave pushed him. Jung's eyes followed the downward slope of the drawing as he thought about how the water lowered as the wave moved past him and got weaker. Jung picked up the pencil and circled the top crest of Pauli's drawing.

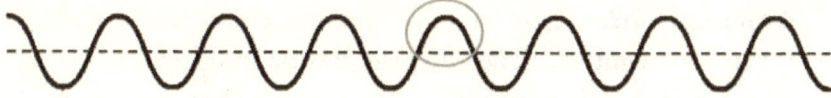

"Incorrect. A wave is a cycle. One single wavelength begins and ends at the same points in the cycle." Pauli drew two vertical lines.

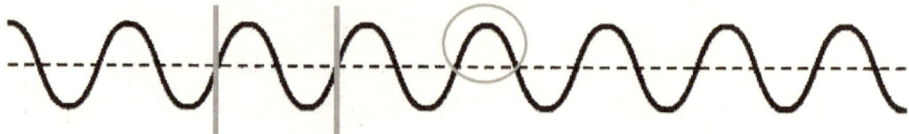

"This is one wavelength." He said.

"Okay, one wavelength." Jung said with a small sigh. "Dr. Pauli, may I ask why any of this matters? I feel like this has nothing to do with the parts of physics I'm interested in." Jung admitted.

"And what part of physics are you interested in?" Pauli challenged abrasively.

"I do appreciate your time and effort, it's just that this is the first of four forces, and I am starting to get lost. I am most interested in the weird things that are happening at the quantum level."

To Jung's surprise, Pauli smiled. "Fear not Dr. Jung for I have deceived you. Once I finish describing the electromagnetic force, I will be able to explain the remaining three forces in one minute each."

Dear Reader,

You do not need to remember all the intricacies that Pauli is describing. Future chapters will revisit these fundamentals after learning other concepts allowing you to see from a new perspective.

You should be gaining an intuitive understanding of what is actually happening in reality versus being able to recognize terms on a multiple-choice exam. There will be no exam.

"That's relieving." Jung said.

"What I am describing now is how waves work. If you are interested in the quirky quantum rules, then you must understand how waves work. And why is that you might ask. The two main theories of quantum physics are all based in vibrating strings or fields of vibrations. These vibrations are literally waves so you must understand how waves work if you want to understand quantum anything!"

"Fields of waves." Jung mumbled, confused at what that might possibly mean.

"A wave is just a shape. What we are truly interested in is the energy in reality being represented by the wave on paper."

"A wave is the shape of pure energy?" Jung was astonished.

"Every wave has two types of energy: frequency and amplitude. Let me explain what frequency is." He added two more waves.

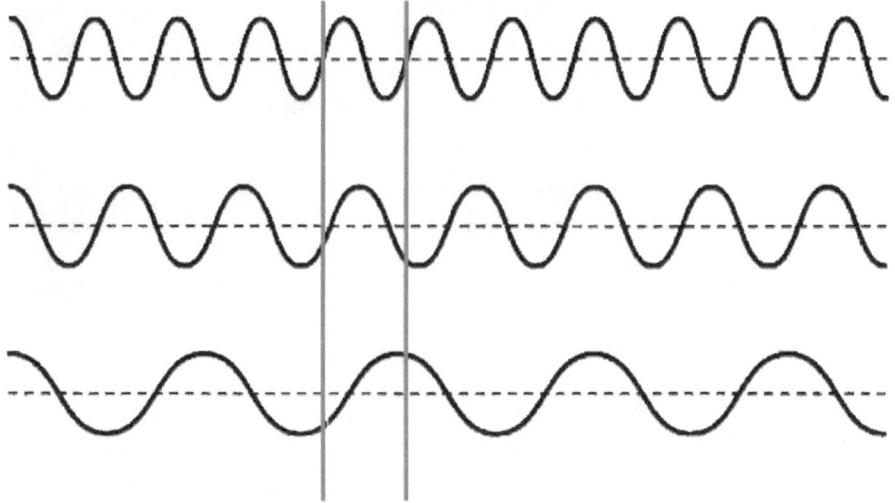

"Why did you extend the vertical line through the other waves? They only match the wavelength of the middle one."

"That is exactly what we need to discuss. These vertical lines represent a time interval of one second. Each wave has a different number of cycles during that one second."

"Okay." Jung followed.

Pauli pointed to the top wave and then the bottom, saying, "this wave is high frequency and this is low frequency. Which of these waves has the most energy?"

Jung immediately wanted to say high frequency because 'high' seems more than 'low' but he reminded himself he really wanted to understand what Pauli was describing so he studied the drawing. "More waves during the same period of time," he mumbled.

If I was standing in the ocean, Jung thought, the low frequency wave seems normal, I would just bob up and down, but the high frequency wave would hit me repeatedly until I fell back. "The high frequency must have more energy."

"Correct again Dr. Jung. For electromagnetism, this change in frequency is what controls whether the energy is visible light, microwaves, radio waves, or something else."

"And the frequency is just a measurement of the strength?" Jung said, a little confused.

"That is exactly *not* what I just said, are you paying attention?" Pauli said, startling Jung. Luckily for Jung, Pauli didn't wait for him to answer. "I knew I should have started with amplitude. Notice in these drawings that the top and bottom of each wave are at the same height. If we add amplitude, then the waves would get taller and the effects would be stronger." Pauli clarified as he drew another set of waves.

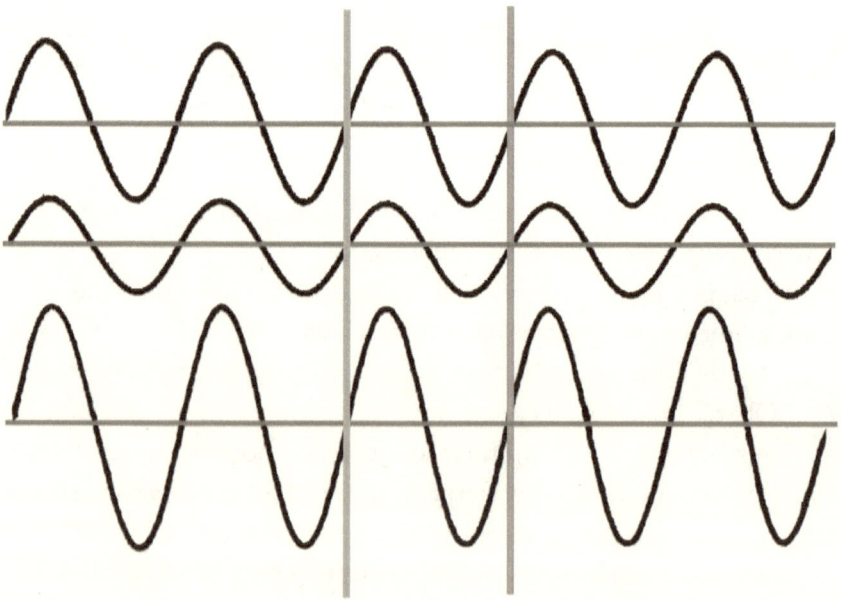

"Notice," Pauli said, "that each wavelength is the same. That means it's the same frequency; it's all the same type of EM energy."

"I think I understand." Jung imagined himself at the beach again. "The frequency of the waves is determined by how often the top of each wave hits me, but the amplitude is referring to if each wave is half a meter tall or ten meters tall." Jung reasoned.

"Yes. The tidal wave versus the shallow wave would be the strength of that visible light, x-ray, or whatever." Pauli explained.

"I get it." Jung thought aloud. "If I see sunlight early in the morning compared to sunlight at noon, it is EM energy with the same wave-*length-*"

"Frequency," Pauli interjected, but Jung continued, without missing a beat.

"Because it is all visible light, but the morning light has a low wave *height-*"

"Amplitude." Pauli interrupted again.

"Whereas at noon the wave-height is high like a tidal wave because there is more energy."

"Now you've got it," Pauli encouraged. "To say it differently, our eyes can only see electromagnetic energy that is in a certain frequency range-"

"Wavelength." Jung interrupted, jumping at the opportunity to return Pauli's banter.

"And has enough amplitude." Pauli said.

"Strength." Jung understood.

"If it doesn't have enough amplitude then we can't see because the light is too dim. If the light gains or loses frequency, then it shifts into a type of energy we can't see because our eyes only see a certain frequency. Let me draw a little diagram."

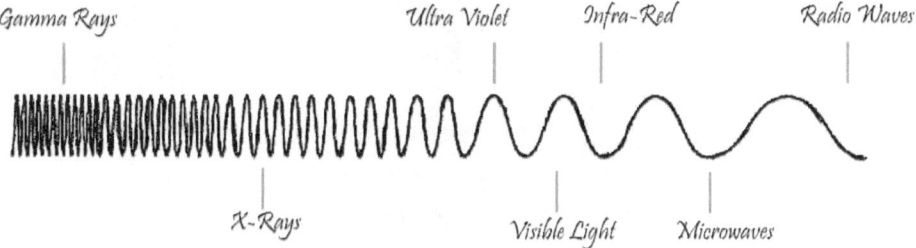

"Notice the amplitude is the same the whole way up and down this wave."

"Right, so the EM energy is the same strength, but the frequency is changing."

"When EM energy's frequency falls in this range," Pauli said pointing to a spot he marked as 'Visible Light,' "we can see it. We *see* different frequencies."

"We can see frequencies?"

"Don't confuse yourself." Pauli snapped. "We see different colors. Each color is EM energy at a slightly different frequency. Now look at this spot." Pauli pointed to ultraviolet light, just outside the parameters of visible light. "Honeybees can see ultraviolet EM energy even though we can't. To them it is just a different color." Pauli explained. [FACT 41]

"Fascinating." Jung said as he marveled at this spectrum of energy that he unknowingly interacted with his entire life.

"Here's one more fact to help you understand. Look at this point on the EM spectrum." Pauli pointed.

"Mic-ro-wave." Jung read aloud.

"Yes. Microwave energy can be used to cook your food but it can also be used like a radio to send signals wirelessly over great distances." [FACT 42]

"All that is the same force; I had no idea. This is very interesting, but you said this same energy is what holds atoms and molecules together? It seems strange that light is holding particles together." Jung stated.

"Let me clarify something. When we are talking about visible light or any EM energy in a certain frequency it requires an energy source. The negative charged electron and positively charge proton do not require an outside energy source because they are continuously interacting. They are simply responding to each other via the field that is automatically produced by virtue of their electric charges."

"That's a lot to take in," Jung paused for a moment. "So which type of EM energy is holding the atom together?"

"The protons' and electrons' interaction is continuous, so there is no specific frequency. It is a field." Pauli explained.

"The magnetic field." Jung figured. "One question before we move on. At the scale of human reality, a positive and negative magnet would attract until they touch, but at the subatomic level the negative electron actually floats around the proton. If they are like magnets, why don't they touch?"

"The two magnets take the path of least resistance to end up in their ground state. In human reality once magnets are touching, they have reached their ground state. That means they are existing in the way that uses the least amount of energy to maintain its natural existence."

"I don't understand." Jung admitted.

"It seems counterintuitive, but when the electron is in its orbital ring it is using less energy than if it was to bind directly to the proton." Pauli explained. "When an electron is in an orbital ring it is in its ground state."

"Okay, I can understand the electron wants to use the least amount of energy necessary but how could moving in orbit be the least amount of energy?" Jung asked.

"This is one of the questions that led to the initial exploration of quantum physics! The hydrogen atom humbled us all. The answer is you are still thinking of an electron like a planet orbiting the sun. In reality, it is incorrect to describe an electron as 'in orbit' at all. For a planet in orbit, we can calculate its exact location and its exact momentum. In quantum physics the more accurately we know a particle's exact position the less accurately we know its momentum and vice versa." [FACT 43]

"Each particle's momentum and physical location are somehow balanced with each other so that we can only know one." Jung stated.

"Furthermore, when we work out the math to figure out where the electron is or what its momentum is, all we can definitively come up with is its most probable location or momentum. Meaning the electron could be anywhere on the orbital plane and we cannot calculate where it is. Instead, we calculate *how likely* it is to be in any one spot."

"Probability." Jung thought aloud.

"If you want to study quantum physics you should become comfortable with things like probability, which, again, are mathematically described as wave functions, which is why explaining the EM force like this is so useful."

"I think I've confused myself again. You described the electron as a force field but now you're saying it could be in any location on the orbital, which insinuates that it is not surrounding the whole atom." Jung said, not sure how to even formulate a question through his frustration.

"You aren't confused at all; you have found particle-wave duality." Pauli knew Jung was discouraged so he said, "the best that most of us can hope to achieve in physics is simply to misunderstand at a deeper level. I will tell you all about particle-wave duality later. First, the fundamental forces. You just learned about the first one; electromagnetism."

Jung's head was spinning. How could all of that be just one force? He took a minute to think back over the conversation. There was a lot of talk about the way electromagnetism expresses itself, but the original question was what binds electrons to the nucleus in a single atom and the answer was; electrons and protons are like tiny magnets, bound together by a magnetic force because of their electric charges. That explains protons and electrons, Jung thought, but what about that third, neutral, particle.

"Well then," Jung continued with a renewed sense of confidence, "now the question is, how do neutrons bind to protons in an atom?"

"Very good!" Pauli laughed, wagging his finger while giving a gentle nod of approval. "Let's discuss the strong and weak nuclear forces. As you noticed, we discussed how electrons bind to the nucleus of an atom but we didn't discuss how the protons and neutrons of an atom are bound together. The two nuclear forces are in constant opposition to each other."

"Two of the forces are related." Jung understood.

"The strong force holds together neutrons and protons creating the nucleus of every single atom in existence."

"Okay, the strong force holds protons and neutrons together. What else does it do?" Jung asked, challenging Pauli to stay under a minute as he said he could.

"As its name implies, it is the strongest of the four fundamental forces. Do you remember what protons and neutrons are made?"

"No." Jung did not even try to guess.

"Quarks." Pauli continued, "the strong force is like a rubber band that holds those quarks together. Here is where it gets interesting, this rubber band holds the up-quarks and down-quarks together in a single proton or neutron while simultaneously holding together the up and down quarks of neighboring protons and neutrons which is how protons and neutrons combine into one nucleus." Pauli paused.

"So, it would be like saying molecules are held together because of something that is holding its atoms together." Jung imagined.

"No, no, no. Your scale is off. Up quarks and down quarks make up protons and neutrons. Protons and neutrons make up atoms." Pauli said as the two men talked past each other.

Jung realized the miscommunication and said, "Yes, I understand the scale, I am just using atoms and molecules as an analogy where the two molecules are held together because of the force holding the smaller atoms together."

"I suppose that's an accurate analogy." Pauli said. "But it's dreadfully inaccurate."

"The nucleus of an atom is held together indirectly." Jung surmised. "The rubber-band-like strong-force holds the quarks together and simultaneously holds the protons and neutrons together. I suppose I understand all that but why does the strong force act like a rubber band at all?"

"By continuously changing the quarks colors with perfect harmonious balance, in a way that maintains the entire proton or neutron." Pauli explained.

Jung's eyes glazed over and he stared at Pauli blankly as if all his thoughts had faded out of his mind.

"That is more than you need to know today though. We will discuss the force carrying particles and then you will understand what I'm talking about."

"So, the strong force acts on the quarks that make up the protons and neutrons and some interaction in there is what binds everything together. Alright, simple enough, I guess." Jung said.

"To go along with the strong force is the weak nuclear force which pushes those same neutrons and protons apart. If you have heard of radioactive decay, this is that. The strong force holds together protons and neutrons while the weak force pulls them apart. These two forces are constantly fighting but as you look around our reality, which one wins?"

"Well, the strong force." Jung said just because of its name, before putting it all together. "Otherwise, nothing would exist!"

"Precisely! Everything in human reality is made of stable atoms which stay together more than they are pulled apart, therefore the pulling-together-force is the strong force and the pushing-apart-force is the weak one."

"And we call them 'nuclear' because they deal with the nuclei of atoms!" Jung exclaimed as he made his final realization. "We have two nuclear forces and the EM force. What is the fourth fundamental force?" Jung asked.

"Gravity."

"Of course." Jung said.

"You may think you are familiar with it because Newton's apple fell to the ground next to him. You may know this force affects anything that has mass. Simply put, things with any physical substance are attracted to each other. The full truth is much more complicated than that, of course!"

Pauli continued, "Gravity is a force that actually warps space itself. So, Euclid's lines are not actually straight when matter is present. You see the edge of your desk; it looks straight to us. In fact, it exists in space that itself, is curved *because of* the mass near it!" [FACT 44]

"Space itself is curved?" Jung said trying to imagine what that meant.

"We will hold that conversation for General Relativity. Here is the problem with gravity. When we are dealing with fundamental particles, gravity is not nearly as strong as it should be. So where did gravity go?" Pauli asked rhetorically.

"Gravity doesn't exist at the quantum level?" Jung pondered, thinking of how gravity affects planets but not particles. "I think I understand now. Even though similarities appear between our solar system and atoms, there are totally different laws of physics that govern them."

"There you have it!" Pauli exclaimed. "All four fundamental forces along with an explanation of what matter is from the cosmic scale right down to the quantum level."

"So, there are four forces that affect everything." Jung thought for a moment, "But how do the forces interact with the matter?" [FACT 45]

FACTS

41.   Bees can see an ultraviolet color but they cannot see red. This difference in sight allows bees a couple of benefits. We see flowers as vibrant arrays of colors but with the ultraviolet spectrum visible, bees see whether or not each flower has pollen in it or if it has already been pollinated. So, when you see a bee brushing itself inside a flower, it was not a random occurrence, the bee is saving time and energy by only going to flowers that still need to be pollinated.

Ultraviolet light shines through clouds better than visible light so when we are experiencing a dreary overcast day bees are still able to see without reduced visibility. [Ref 34]

42.   The microwave oven was not invented until 1945, 13 years after the timeframe in the narrative, so these men did not know that microwaves could heat your food. Microwaves cook our food by emitting bursts of electromagnetic energy as your food rotates around that little glass plate.

Temperature is a measurement of how much vibrating a molecule is doing. The hotter something is, the more its molecules are vibrating. So how does the microwave oven make molecules vibrate?

Water, along with other molecules in our food, are polarized. This means one side of the molecule has a positive charge and the other side negative. Since microwaves are electromagnetic energy, they naturally interact with electrically charged particles. Your microwave shoots a burst of EM energy that makes all the polarized molecules face a certain direction then it quickly shoots another burst that forces them to flip the opposite way. This molecule-flipping causes them to vibrate and therefore gain heat.

Microwaves can also be used exactly like radios. No not your microwave oven, the actual EM frequency, microwaves. This is not the preferred EM frequency for sending transmissions because water vapor in the air can distort the signal thereby disrupting the messages being sent. Also, this type of communication requires a more direct path from transmitter to receiver, so it does not scale up well. [Ref 35]

43.   The Uncertainty Principle, or Heisenberg's Indeterminacy Principle – Quantum particles have many unique properties, some of which are

complimentary. In other words, there are pairs of properties which are grouped together. One such pair is momentum and velocity. The Uncertainty Principle postulates that the more accurately we know one, the less accurately we are able to know the other. German physicist Werner Heisenberg first postulated this theory in 1927.

44.   Mass warps space. This curvature in the fabric of space-time is what we call gravity. Continue reading for an understanding of General and Special Relativity. [Ref 36]

45.   The descriptions of the four fundamental forces in this narrative are scientifically accurate. [Ref 37]

Chapter 17

Coincidence or Synchronicity?

Pauli swelled with pride as his pupil excelled. Between last week's lecture on matter making up the stuff of the universe and the fundamental forces that Pauli explained today, it seemed Jung had no issues grasping the core concepts of physics. After returning from a break Pauli made himself comfortable on the chase lounge once again.

"How do forces and matter interact? Very good question." Pauli began. "The forces are actually carried by different fundamental particles."

"There are other particles?" Jung exclaimed. "And they *carry* the forces? I thought you already discussed all the particles."

"Well, no," stated Pauli. "We only discussed the fundamental particles that build up to all the stuff we see in human reality; the up-quark, down-quark, and electron." Pauli started thinking about the force-carrying particles and said, "actually, the forces might not all be carried by particles. We know the photon carries the EM force, gluons carry the strong force, bosons carry the weak force but then there is gravity. We only guess that the graviton particle carries it, but gravity becomes so weak at this quantum scale." Paul explained.

"Wait, why is electricity not one of these forces? You said the electron is a fundamental particle which carries electricity. It seems exactly like one of these forces." Jung challenged.

"The electron is a fundamental particle because it cannot be broken down into any smaller physical parts."

"Are you saying it has non-physical parts?"

"Yes," Pauli said mystically, "but that is not a conversation for today. Just because it is a fundamental particle does not mean it is a force-carrier or matter-builder." He explained. "A photon literally is light. The electron is not electricity. Electricity is generated by flowing electrons. When we use light, we use up photons. When we use electricity, we don't use up electrons, we use up the energy that is moving the electrons."

"So, when something runs out of electricity the total number of electrons hasn't changed?" Jung figured.

"That's right. The metal that conducts the electricity is comprised of atoms that each have one lone electron in their outermost orbital."

Jung visualized Pauli's drawing of an atom with one single electron in its outermost shell.

"When we use electricity," Pauli continued, "we are using the energy moving the electrons, not the electrons themselves"

"This is just like when you described waves at the ocean." Jung understood, "I thought the wave hitting me was due to the water but really the wave is the energy the water possesses."

"Exactly. When the wave hits you at the beach or when electrons flow through a wire, the energy moving the thing is what we are talking about. This is what allows for alternating current."

"What is alternating current?" Jung asked.

"Our electronic devices use one of two types of electricity: direct or alternating current. In direct current all the electrons continuously flow in the same direction down the wire and through the device being powered, then back to the power source. In alternating current, the electrons' direction is oscillating."

"Meaning the electric current changes directions?"

"Precisely," said Pauli. "The feud between Edison and Tesla was about this very subject. Ultimately Tesla's alternating current won because of its scalability, but I digress." Pauli cleared his throat before continuing. "Whenever we have flowing electrons, we also must have an electromagnetic field. Together, they are both the inseparable force of electro-magnetism."

"But if electricity creates a magnetic field, then it seems that electricity by itself is more fundamental than the EM force. It seems like electro-magnetism should be divided into two forces so why is it just one?" Jung asked.

"You're thinking like a philosopher. We use electricity differently from magnets, sure. But in physics they are fundamentally one thing that cannot be divided. If electricity is flowing then there is a magnetic field because they are one, no exceptions."

"What about the Earth's magnetic field. Where is all the electricity from that?" Jung challenged.

"Excellent question," said Pauli. "As we speak there is more electricity flowing around Earth's core than the worst thunderstorm you could possibly imagine."

"Oh," Jung relented as he imagined lightning cracking through the densest part of the Earth's core.

"Now think about EM at the scale of particles. Electromagnetism binds electrons to protons, not because they are magnets but because they are electrically charged, and that electric charge reacts to the magnetic field; it is fundamental for our existence." Pauli explained.

"What particle carries EM energy again?" Jung asked.

"The photon."

"Well how do photons relate to electrons?" Jung asked, hoping to gain a better understanding of electromagnetism.

"Good thinking." Pauli started, "electrons can absorb or shed photons."

"How does that work?"

"The photoelectric interaction. It's when a photon is absorbed or emitted by an electron which causes the electron to jump from, or to, a higher energy level."

"Oh boy." Jung said, feeling like he bit off more than he could chew.

"In this case, don't think of the photon as a particle." Pauli instructed. "It is the photon's energy that can be absorbed by the electron at which point the photon no longer exists."

"Because its energy was absorbed." Jung said.

"A photon by itself is electromagnetism. An electron by itself is not electricity; that is why one is a fundamental force and the other is not. The electron's naturally occurring property, specifically its negative charge, is why it is one manifestation of electromagnetism."

"But why is an electron negatively charged, what if it didn't have a charge at all?" Jung struggled.

Pauli was frustrated and blurted out, "if your grandmother had wheels, she'd be a bicycle!"

"What?" Jung laughed.

"I am not a philosopher," Pauli explained, "I don't tell you why something is, I only define what it is and in this case an electron's inherent property is that it has a negative charge. If an electron had a neutral charge it would not be an electron!"

Jung thought for a second and said, "I feel there are aspects to physics I don't fully understand, but I can accept what you are saying as truth so we can continue the conversation."

"That might be the best evidence of your genius. In fact, all physicists must take a leap of faith with one understanding or another. Many physicists who have predicted correct theories die of old age before their theories are proven right." Pauli said.

"Can I tell you," Jung said, "what I am most curious about is the idea of an invisible field of energy."

"Ah, yes. The concept of fields of energy will be a much deeper conversation, another time." Pauli said.

"In that case," Jung said. "I will ask; where do those force particles come from?"

"Ha! I must say I am a bit impressed, Herr Jung! You are a master of your craft and in just two afternoons you have learned all the basic concepts of classical physics right up to knocking on the door of quantum mechanics! I still don't understand why you have such an interest in physics. It seems to me our disciplines are as different as could be." Pauli said.

"I have a feeling they may not be so different after all. I am studying a theory I call synchronicity. The thing is, I cannot understand how it could possibly be…" Jung paused to find the correct word, "…real." He finished his sentence with a shrug.

"What is synchronicity?" Pauli asked curiously.

"When an observer sees two events that are not related but the observer can assign meaning which makes it seem as if the events are connected." Jung said.

"Give an example." Pauli said for clarification.

"When the idea of a certain individual pops in your mind from nowhere and a moment later the telephone rings. When you pick up, of course it is that person on the other end." [FACT 46]

"I am skeptical of this. Perhaps when you hear the phone ring, your mind runs through a list of people who may be calling, and your mind automatically thinks of the most plausible person, given the circumstances of that moment." Pauli challenged.

"What about when you think of the person *before* the phone rings? Here is another example that cannot so easily be explained away." Jung said gracefully. "I once had a client who was so logical, he could not see any value in analyzing his dreams. I was intent on showing him that his dreams were trying to tell him something; they weren't just random images.

"During one of his counseling sessions, he described a dream about a scarab beetle. As I listened, I was distracted by a rapping on the window. When I opened it a strange shiny beetle, which certainly did not belong in Switzerland, flew in. I caught it easily as if it wanted me to. When I presented it to my client, he was so shocked that it broke down his logical barriers." Jung explained, thinking back to Rupert all those years ago.

"It seems an odd coincidence, yet it does seem to be just a coincidence." Pauli said.

"A synchronicity is a coincidence with meaning." Jung explained. "What if there is more to a common coincidence than meets the eye? Let's breakdown this synchronicity. There were two parts: the dream of a beetle and the real beetle. The dream beetle is a symbol cast from the subconscious mind. It should have no connection to physical reality unless the dreamer decides to act upon his dream later.

"The real beetle could have tapped on any window. It could have tapped on no windows at all and yet it tapped on mine, repeatedly, at the exact moment my patient was describing something just like it. It was only him and myself who could see the connection between the two events. The beetle could have flown past someone outside who would not have noticed it at all."

"So, the dream did not cause the beetle and the beetle did not cause the dream." Pauli said.

"And yet, there it was, the beetle synchronicity." Jung remembered.

"Yes, that is quite odd." Pauli agreed. "It's just hard to say where pure chance ends and something more begins."

"Have you never experienced any coincidence you thought was so odd, you felt there must be something more at work?" Jung asked.

Pauli thought for a moment and then gave a loud sigh. "In fact, I have. It is now to the point that my colleagues no longer allow me in the vicinity of any experiments. It's like a curse. Devices, computers, even everyday objects break in my presence. I suppose it has happened so many times that it feels like more than a coincidence." [FACT 47]

"Things just break?" Jung asked.

"Most recently, I was driving from Zurich to a conference in Ulm. I received the invitation late but a friend of mine convinced me to go because he was born there. I thought it would be a good opportunity to spend some time with my wife outside of our normal routine, so I tried to book a train. On short notice I couldn't get a connection, so I was forced to drive early the morning of the event. I am not a morning person and as I arose, I stubbed my toe on the bed post which set me off. I was having one of those mornings where nothing went right.

"We stopped for a quick breakfast along the way and when we tried to get back on the road, the car wouldn't start. I tried everything I could think of, all the while my wife berating me as if I was doing something wrong. I contacted a mechanic, but he couldn't find anything wrong, so he went to turn it on and it started without hesitation."

"Your car broke but was perfectly fine in the presence of a mechanic." [FACT 48]

"The Pauli Effect was to blame, yet again."

"The Pauli Effect?" Jung asked.

"It's so prevalent they coined a name." [FACT 49]

"How did this 'Pauli Effect' begin?"

Pauli sighed. "We were conducting an experiment at the University and everything seemed to turn on like normal but as we went to run the test the computing machines all turned off. We worked late into the evening looking for the cause but found nothing. The next day my colleagues made no changes and ran the experiment just fine without me.

"A day or two later they ran another instance of the same experiment. I was running late and as I walked in, it shut down without any obvious cause. I decided to sit there and figure out what the problem was. After finding nothing wrong with the machines, I left with my colleagues; but they had formed a hypothesis that somehow my presence was to blame. They doubled back to start the machines again. Without me there, they worked without issue."

"So, you are cursed." Jung said, almost with a chuckle.

"It feels more like I *am* the curse." Pauli said dramatically. "But what exactly are you suggesting is causing this type of coincidence?"

"Could these things be more than coincidence? Could there be a quantum explanation to all this?" Jung asked.

"Perhaps you are right to explore this new field of quantum physics but as far as we can tell, it is all just coincidence. I think you are searching for something much deeper like a scientific test to prove God. How could anyone succeed in proving God?" Pauli asked rhetorically.

"It is our job as scientists to gain a deeper understanding and if we stop trying just because we are scared of defining some aspect of God then we are forsaking our profession." Jung proclaimed. "In truth I believe aspects of God should be perfectly explanatory. Either way, this is why I am so interested in physics. These events are clearly not related yet somehow, they *feel* connected."

"Feelings are difficult to quantify." Pauli stated.

Jung drove the conversation forward. "Each part of a synchronicity has a logical explanation. The odd thing is that we notice a connection because the two things appear at the same moment in time and space. It seems like the synchronicity is greater than the sum of its parts. Perhaps…" Jung started to say.

"Perhaps somewhere in particle physics there is an *actual* connection." Pauli replied, finishing the thought.

"Or in whatever science explains the realm that creates particles which I am still waiting for you to describe." Jung added with a cheeky smirk.

"These synchronous events do not *cause* one another," Pauli thought aloud, "nor were they caused by the same thing, or even by a chain reaction of events that originated from the same source. It is odd, I'll give you that, because it seems to us, the conscious observers, that there is some connection."

"To me," Jung said, "it would seem as though there's an unseen connection that only exists in some ethereal realm which must be beyond our human reality." He explained the best he could in the moment. Jung preferred to correspond by writing letters because he could take time to collect and mature his thoughts before sharing them.

"Spooky action at a distance…" Pauli mumbled before trailing off.

"The thing about synchronicities is that without the observer there would be no connection at all." Jung pondered aloud.

"Particle-waveform collapse…" Pauli muttered.

"Sorry?" Jung said.

"The human making sense of it all is causing quite a conundrum for modern physicists."

"How is that?"

"It is not possible to observe particles without changing them. It is a fundamental issue that makes studying them so difficult." Pauli explained.

"You are saying that the act of observing an experiment actually changes the results?" Jung said.

"Yes, but not because of human consciousness. The tools we use to measure any single particle eradicates it in the process, so we have no way to observe a single particle while it is in superposition." Pauli explained.

"Super what?" Jung asked.

"Ah yes, we have not yet discussed all the strange behavior of particles. All I mean to say about this observation conundrum is that to actually prove anything we must observe what is happening. How can we get any data if we don't know what's there? The problem is that in order to actually see what a particle is doing; we change its properties."

"You are saying, the only way to see what a particle is and does, is to use tools which change the particle in the process?" Jung clarified.

"Essentially yes."

"So, a moment before measurement, the particle could have been something totally different?"

"Not totally different, but yes, different; this is the inherent problem with the observer. By measuring the particles, we may be changing them." Pauli sat and thought in silence for a few moments. "I appreciate your quest to find the deepest truths of our human reality."

"You have taught me much and yet I am desperate to find out more." Jung said patiently, sensing the lecture was ending.

"There are many thoughts in my mind. The trouble is how to describe them to you in a way you will understand. You are smart, but raw intelligence does not equate to specialized knowledge gained over years of experience in a field. I am used to giving lectures to individuals who already have a deep understanding of physics so I must collect my thoughts for this unique lecture." Pauli said. "I like you, Herr Jung. I have a schedule to keep but I look forward to our next session; both for myself and for your physics."

"Likewise, Herr Pauli, I think we will learn many things from each other." Jung said.

The men ended their discussion and scheduled another meeting for the following week.

FACTS

46.   Even in the days before caller ID, people would often experience a phenomenon where the idea of a person would pop into their mind even *before* their phone began ringing. When they picked up, they found that person on the other end. It was like the act of dialing a phone with the intent of reaching a certain person, created a psychic connection that called out to the intended receiver. Are you too young to remember this? Then ask one of the elders in your life, they have this knowledge. [Ref. A person in your life.]

47.   There is a phenomenon where things stop working in a certain person's presence. Wolfgang experienced this to the point that his colleagues banned him from being near experiments. As time went on that sentiment was quelled as nonsense but for Wolfgang Pauli it was all too real. [Ref 39, 40]

A similar phenomenon reported by many people is called Street Light Interference. Reportedly, when specific people walk by a certain streetlight it always turns off the moment they pass.

48.   Has your car ever made a funny noise or had some issue and when you took it to the mechanic, they couldn't find anything wrong? What about your computer acting up, except in the presence of the IT guy? For Pauli this event did occur but under slightly different circumstances as described in the narrative. It was 1934, on his honeymoon with his second wife. The car inexplicitly broke down but worked as normal in front of the mechanic. [Ref 39]

49.   The phenomenon of Pauli's mere presence causing things to malfunction or break was dubbed the Pauli Effect. The name stuck and the Pauli Effect is now a controversial idea. How could this constant coincidence be anything more than just that? [Ref 39, 41]

Chapter 18

Multi-Dimensional Space-Time

"People like us, who believe in physics, know that the
distinction between past, present and future
is only a stubbornly persistent illusion."

-Albert Einstein

One Week Later

Pauli arrived at Jung's office first thing in the morning. His usual appointment was in the early afternoon, but he rescheduled today because of a conflicting meeting in the physics department. Jung asked Pauli about the troubles in his marriage. Pauli mostly described issues he could legitimately blame on his wife. Jung continued pressing for more. Pauli started to trudge through the cringey memories of his own wrongdoings and finally talked through some of the things he did wrong in the relationship. It seemed like he was close to a breakthrough, but Jung felt like Pauli was holding something back.

"Doctor Pauli, something seems different about today's session. Are you still taking this seriously?" Jung asked in a moment of frustration.

"I assure you I am taking this counseling seriously Doctor Jung! I do not often wake so early and when I do, it is only for a very good reason." Pauli snarled.

Jung lightened his tone and offered, "then perhaps I can buy you a coffee."

After the counseling, the pair walked through a park towards a small café. The green island in the heart of Zurich had a canal on one side and trees on the other which made it feel as though they were out in the wilds of the countryside.

"I've been having a dream." Pauli admitted to Jung as they walked.

"Oh? For how long?"

"It started about a month ago. Now I dream it almost every night. It has even started to wake me up!"

"Can you describe it?" Jung requested.

"It starts as an actual memory where I am sitting in class at my preparation school, Döblinger Gymnasium, back in Vienna, I must have been sixteen at the time. I was listening to a lecture but my mind was wandering as it often did. I don't remember being hot but I do remember the window being open and a refreshing breeze crossing my desk.

"In the moment, I was staring out the window, contemplating Einstein's theory of General Relativity and some of its implications. Up until this point the dream was accurate to my memory but that quickly changes. "In the dream the teacher scolded me for being distracted and to my surprise, when I looked back at the chalkboard, it was filled with formulas about General Relativity. I can't remember what topic was being taught in real life but it most certainly was not General Relativity seeing as this was high school and the theory was brand new at the time.

"I don't know why but for some reason I felt called to look back out the window even though I knew I shouldn't. I gave in and stared outside knowing I was risking the ire of my teacher. As I stared into the sky a giant dragonfly appeared. It flew directly at me as if it was going to attack.

"It all seems so silly once I wake up and think about it, but during the dream it feels like a very serious situation.

"I close my eyes and swat at the dragonfly but it never comes. When I open my eyes again, I am no longer in the classroom. Instead, I'm standing on the edge of a pond looking out at a large white goose swimming around."

"Is the water in the pond calm or rough?" Jung asked, immediately recognizing the significance of the water symbol.

"It is very calm, even with the goose swimming, the pond is as still as glass, like a mirror reflecting everything that looks at it." Pauli said as something behind Jung caught his eye.

"What happened to the dragonfly?" Jung asked, as he instinctively glanced over his shoulder to see what Pauli had noticed. Jung watched an owl swoop into some shrubs hoping to catch a morning snack as Pauli continued.

"The dragonfly is there too. It is flying directly towards me from the direction of the goose. I feel like it wants to attack me." He paused.

"What else happens in this dream?" Jung asked.

"That's all I can remember."

"When you saw the dragonfly what were your feelings towards it?" Jung asked, trying to get as much information as possible.

"I felt like it hated me. It wanted me to know it hated me but somehow, I know it couldn't harm me unless I let it." Pauli explained.

"What color was the dragonfly?"

"Red."

"Curious," Jung said. "What are your thoughts on what your dream is trying to tell you?"

"I haven't the slightest idea."

DING-DING, DING-DING.

The two men stopped abruptly on the edge of the sidewalk and looked to their left. A streetcar turned onto the road and clanked by. They reached the edge of the park and were reentering the bustling cityscape. The Swiss National Museum's medieval-style tower stood proudly like a castle in the bright morning sun behind them. They continued across the street, passed the busy train station and turned towards the café which sat at the base of a steep hill.

Sitting proudly on top of the hill was the University of Zurich, where something green caught Jung's eye. It was the squared dome atop the tower at the center of the small campus whose green brass was brightly reflecting the morning light. Jung paused, trying to decide if it had always been such a vibrant green.

"Okay," Jung said, "in that case let's break down the different elements in the dream and figure out what each could symbolize. There is your school and the pond and then there is the teacher, the dragonfly, and the goose. Am I missing anything?"

"Those would certainly be good places to start."

"Very well. Allow me to reference my notes for these symbols and we will discuss this dream at your next session." Jung said as they arrived at the café.

The men were served their coffees, and Pauli immediately began the physics lecture. "Do you remember my comment on lines not actually being straight? I believe I used your desk as an example."

"Vaguely." Jung said, already on the edge of his seat. This was just the type of thing he was hoping to discuss. Something that was scientifically accurate yet clearly did not fit into the experiences we see in human reality.

"Well, we were discussing gravity. Today, I will clarify as we discuss General Relativity. This theory defines the relationship of space-time."

"Space and time." Jung said lightly.

"No." Pauli said definitively before cracking a smile, as if he got some sort of enjoyment by telling Jung he was wrong. "Spacetime. For this, you will need to forget what you think you know of the two."

"Spacetime?" Jung thought aloud. "There is outer space, which I guess is the same thing as the space we exist in here on Earth; we just have so much other matter to interact with that it seems different. How complicated could space be?" Jung joked. "And time? Time is a universal constant, is it not? So, it is experienced by everyone the same?"

"In human reality, time is experienced the same by all individuals but that does not mean it is constant." Pauli said mysteriously. "General Relativity changed our deepest understandings. A dear friend of mine…actually you may know him - Albert Einstein, he works right here at the University." Pauli pointed up the hill towards the university's green roof. He postulated the theory of General Relativity."

"I did indeed meet Herr Einstein; I had the pleasure of hosting him for dinner one evening." Jung replied. [FACT 50]

"In human reality, we experience three dimensions of space and a dimension of time. Four dimensions that explain our everyday experiences."

"So, we live in a four-dimensional space-time reality; how does that work?" Jung asked.

"Don't think of it as four equal dimensions. Three spatial dimensions can be grouped together but time must be separate, at least for now. You see time is *relative* to the spatial dimensions. Before I lose you in that thought, let's talk about the world if you existed as a two-dimensional creature. You would experience time just as you do now, even without the third spatial dimension. There would, however, be interactions between your two-dimensional world and the three dimensional one you truly live in, which would be quite interesting." Pauli said.

"Interesting how?" Asked Jung.

"First let us picture your two-dimensional world. Take this napkin for example." Pauli moved the napkin to the center of the table and drew a triangle with some shapes inside it.

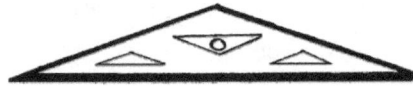

"Consider this napkin to be perfectly flat, literally two dimensions with no height whatsoever. Imagine you are a two-dimensional creature who lives here. This napkin is like your Earth and beyond it is outer space."

"Okay, so I am confined to the napkin and outer space is all around it." Jung said waving his hand above the napkin.

"Hang on. You live in two dimensions. There is no above or below. Your outer space is just an extension of the same plane the napkin is in."

"I see." Jung began, "my outer space is only in the plane the napkin is in but extended out passed it. What would two-dimensional me see?" Jung asked.

"I can show you! Move your head down. Stop when your eyes are just above the table." Pauli instructed.

Jung scooted his chair back and bent his head down, focusing on the triangle drawn on the napkin. As he lowered his head the triangle morphed until it was more or less just a black mass of ink.

[*Special*][*You can try this too. Tilt your book forward to line up your eye above the bottom of the page and look at the triangle.*]

As he moved down Jung said, "the triangle is distorted. I can't see its actual shape or depth."

"Notice the distortion," Pauli instructed. "The triangle looks more like a line than a shape, especially at the edges. It's not totally a line because you are still three dimensional, but the point I'm making is the distortion. If you were truly two dimensional, all you would see is lines or dots no matter what you were looking at," he explained. "Now, what if a three-dimensional object was to move through your single plane of existence?"

They sat up in their chairs as Pauli held the napkin flat in his open palm about eye level. He picked up the small coffee spoon and held the tip of it beside the napkin to demonstrate the interaction. "The spoon is intersecting with the outer space of your two-dimensional existence. As the spoon is lowered through your plane, what is the first thing your 2-dimensional self would see?"

Jung imagined his single plane of existence and did his best to forget about above and below. "I suppose I would see the very tip of the spoon appear from thin air. It would appear as a small line or even just a dot at first."

"Correct, and you would only see the section of spoon that was intersecting with your plane of existence." Pauli explained. "As the spoon moves through, the small dot would grow into a line and depending on your perspective you might see that it's curved." Pauli lowered the spoon as he spoke. "As the bowl of the spoon continued lower, you would see the crescent-shaped line grow to its maximum size before it starts to get smaller, until only a cross section of the spoon's handle was left in your plane."

"And once the handle was through, it would pop out of existence altogether!" Jung interjected. "This is a very interesting perspective, but it doesn't change my understanding of space. I mean, we don't have random objects appearing in our existence from some higher dimension."

Pauli remained silent glaring at Jung with a sly grin. "Do we?" Jung asked, with a rather alarmed tone.

"We just might," Pauli said. "But this would be particles and even sub-quantum particles; nothing we would see at the scale of human reality. Let me point out the shapes inside the triangle."

"Yes, what are those?" Jung asked.

"That could be one of the shape's organs. From this two-dimensional perspective no one would be able to see the shapes unless they cut their friend open. The interesting part is that from our three-dimensional existence we can see what is inside our two-dimensional friend."

"That is interesting." Jung said as if he was trailing off into a thought.

"Therefore, a creature living in the fourth dimension could actually see inside each of us just as we are seeing inside the triangle."

Jung's mind raced as he thought about life at higher dimensions.

"But I digress, we are discussing relativity."

"Right. Time is connected to space, or they are the same somehow." Jung said, hoping to get through this as quickly as possible so they could get to the part of the lecture where particles spontaneously appear in our reality, presumably from some higher dimension.

"Another way of looking at space-time is to say that right now, as you sit in that chair, you are moving through space-time at the speed of light. You must add the rate of travel through space to the rate of travel through time. That summation will always equal the speed of light." [FACT 51]

"Seeing as how you cannot control how fast you travel through time, the only way for us to manipulate anything is to travel faster through space. When you do, your rate of travel through time actually slows down."

"Can we experience this change?"

"Not directly," said Pauli. "Time will always seem to move at the same rate for each of us, individually. The thing is time only slows down *relative* to the things not moving."

"Ah, I am starting to see where the name comes from, it is a general rule of relativity. So, if I am running past someone who is sitting still then I am traveling slower through time than they are?"

"Technically yes, but the difference would be so incredibly small you would never be able to detect it at that velocity. To actually experience the difference, we would need great distances and a vessel that could travel very, very fast. If you get in a spaceship and travel at almost the speed of light you will see an incredible difference. If you could travel at just one meter per second slower than the speed of light and maintain that speed for one year, it will feel like a normal year to you but it will have been over 12,000 years here on Earth." [FACT 52]

"And all the people on Earth would also experience time as normal because nothing will have changed for them." Jung said.

"Exactly! I should mention, however, in that equation you immediately start and stop at that speed. I removed the time it takes to accelerate and decelerate safely."

"But where does this," Jung tried to find an accurate term, "time-mutation end?"

"Time *dilation*, is the terminology." Pauli corrected.

"Ah, time dilation," Jung said. "What is its limit? What I really mean is; would it be possible to go into the past?"

"To quote my friend, Einstein, 'people like us, who believe in physics, know that the distinction between past, present and future is only a stubbornly persistent illusion.' [Real Quote]. Perhaps the past, present, and future are all happening simultaneously. Also, mathematically speaking, the future and the past have an equal amount of influence on the present moment."

"The future can influence the present?" Jung was astounded.

"It's called causal attraction," Pauli explained. "It is based in a new interpretation of mathematics. Think of dropping a ball from a roof. It is a rubber ball, so it bounces. In this case the Earth itself is the attractor that is causing the future state of the ball."

"The future state of the ball…" Jung repeated to understand the perspective from which Pauli was speaking. "You mean when the ball is no longer a ball?"

"No, no, I mean it will be in a state of rest, simply lying on the ground as opposed to when you drop it and it's in a state of motion," he explained. "To understand this way of thinking you cannot think of cause and effect, remember, in this case the future is influencing the present."

"We have to change our understanding of the flow of time." Jung suggested, though he didn't fully understand what that meant.

"That's right. We experience time like a movie constantly playing but for this I need you to pause that movie so we can analyze each passing moment. After all," Pauli added, "we cannot actually experience the past or future, only the present moment."

Jung imagined stopping time as if his life was a movie on film and he was looking directly at the reel, able to pause any single moment.

"Imagine you are on the roof and you drop a ball," Pauli instructed. "The ball will fall, hit the ground, bounce a few times, and finally come to rest, sitting perfectly still on the ground."

"I see," Jung said, "the ball was in a state of motion but now is in a state of rest."

"Yes, now that you have seen the future, rewind that movie to when the ball was falling."

"Okay."

"Pause your movie while the ball is in midair. Does it look any different now compared to when it was resting on the ground?"

"No."

"The ball itself is the same, what is different is its current state of motion compared to its future state of rest. Tell me, what is causing its future state?"

Jung thought for a moment and said, "gravity."

"The Earth's gravity, very good. We can calculate the exact velocity and acceleration because we understand gravity. In other words, the ball's future state of rest is being attracted into the present moment by the Earth itself."

Jung pondered this new perspective of time. "But how is that any different than just guessing the future based on known trajectories?"

"Excellent question," Pauli sounded very excited. "In that example you saw the ball come to rest before you rewound the movie. Likewise, in causal attraction the future state of the ball has already happened." [FACT 53]

"Whoa," said Jung. "Obviously, that is not how we experience time. Then the question is why did humans evolve to experience time as we do?" Jung asked.

"I appreciate that you always bring the human perspective back into the picture. Sometimes, we physicists explain the universe but forget our place in it." Pauli continued, "The future's influence on the present is a mathematical truth but we have no observations of it actually happening."

That's odd, Jung thought. "Is it possible for math to allow for something that reality cannot actually produce?"

"You are asking whether the future influencing the present could be true in mathematics without being true in reality. We are now discussing concepts where just about anything is possible," Pauli said mystically. "I think, most likely, science is simply missing a piece of information to this great puzzle."

"We evolve to survive, both as individuals and as a species." Jung said as he thought through the implications. "Therefore, it may be safe to assume that threats to our safety only exist in the human-reality that we experience. There are no direct threats to us from the past, future, or from any higher dimensions," he figured.

"I would agree, but this spacetime conundrum goes deeper than you know." Pauli stated, staying focused on the physics.

"What now?" Jung said jovially.

"When you travel close to the speed of light you would experience time the same but as you looked out your window, space itself would appear to warp around you. It would be a very odd thing to see. If you looked ahead to your destination, it would actually appear farther away. If you watched your destination as you traveled, it would seem that you were not getting much closer and then, in an instant, you would have arrived. Everything on either side of you would look as if it had been stretched out. Not to mention blue and red shift." [FACT 54]

"That is because space itself has warped?" Jung asked.

"Correct! If you were able to watch someone traveling at that speed it would look like they are being compressed from front to back and they would actually look slightly rotated." Pauli explained.

"Thinking back to gravity," Jung began, "you said it bends space itself, and now you're telling me that time can warp space. There also seems to be some mystery with gravity because it's so much weaker than the other forces. Is it possible that time and gravity are two expressions of something deeper?" Jung was now connecting the dots like a seasoned physicist.

"Perhaps I should bring you to the physics department, Dr. Jung!" Pauli said. "You are now asking all the same questions we are! It is possible that time being warped *causes* gravity." [FACT 55]

"Absolutely fascinating." Jung said, not knowing what to make of this new suggestion.

"There is one more thing. Mass."

"Mass?"

"Mass is both what feels the effects of gravity and what causes the effects of gravity. Mass is what prevents things from traveling at the speed of light. Light itself; do you remember which force light belongs to?"

"Electromagnetism!" Jung exclaimed, happy to be keeping up.

"Well, the EM force is carried by the photon, which is a fundamental particle with no mass."

"No mass, meaning it can travel at the speed of light." Jung said before realizing the extremely obvious thing he was saying. "HA! The photon *is* light, which travels... at the speed of light! But that means that light itself is not affected by gravity either, right?"

"Good thinking, but actually, no." Pauli explained. "The velocity of the photon gives it mass-like properties with regards to gravity."

"Very odd."

"One more thing!" Pauli exclaimed. "If you could ride a photon and travel at the speed of light you would not experience space at all."

"What does that mean?"

"From the photon's perspective when it is created it instantaneously hits its destination with no perception of a trip at all." Pauli explained. "And here's another odd fact; gravity travels at the speed of light."

"That is odd. Gravity, time and light are all mixed up together." Jung let out, as he let all these new ideas permeate the deepest recesses of his mind.

FACTS

50.   Dr. Jung hosted Albert Einstein for dinner on multiple occasions beginning as early as 1911. Jung described in a letter to Freud that he spent the whole evening discussing physics. [Ref 42]

51.   We are all traveling through space-time at the speed of light and when you change one aspect the other must also change because the total will always equal the speed of light. This distribution can be described mathematically as a vector with the understanding that each dimension is a velocity (because it has a direction). The four-velocity vector equation is $U = \gamma(c, v\_x, v\_y, v\_z)$ where γ is the Lorentz factor. [Ref 43]

52.   The speed of light is 299,792,458 meters per second. If you were able to travel at 299,792,457 meters per second for one year that would create a

difference in perceived time of around 12,000 years (as compared to someone not moving). This equation is just a thought experiment to conceptualize time dilation because you would immediately start and stop, which is not possible. Another interesting fact is when you increase your speed to those extremes your mass would increase exponentially.

It really is all relative. When comparing two reference points on Earth we have already removed the factors of cosmic motion. The Earth is moving around the sun but our whole solar system is also moving through space. If we were to compare one point on Earth with a reference point on some alien planet, then we would have to account for the rate of travel of our planet, solar system and galaxy as compared to theirs. (Not to mention the continued expansion of the universe.)

In other words, both locations would experience time the same but there could be a vast difference in our planet's baseline rate of time if another advanced society's home planet is traveling through the universe at a significantly different speed. [Ref 44]

53.   Causal Attraction, or more commonly Retrocausality, is the idea that an effect can occur before its cause. This does not happen in the macro world we see in human reality; it only has a place explaining some quantum weirdness. [Ref 89, 90, 91]

54.   If you were able to travel close to the speed of light you would see many odd things as you look out your window. The description in the narrative is accurate but also you would experience redshift and blueshift. This is caused by red light's longer wavelength which travels faster than blue light's shorter wavelength (when not in a vacuum).

Therefore, when you travel at almost the speed of light the different wavelengths of light reach your eye at different times. Objective reality is not changing in this example, it is just how you interpret what you are seeing. Your eyes and brain are wired to interpret light based on the speed you normally travel. [Ref 45]

55.   The warping of time might cause gravity. This is probably not accurate, but we need more research. It is mathematically possible. [Ref 46]

Chapter 19

The Hidden Mind of Consciousness

"We are the music makers. And we are
the dreamers of dreams."

-Arthur O'Shaughnessy, Ode

"Now let me ask you a question." Pauli said as the two men
continued their discussion at the café in Zurich. "Your theory on
synchronicity, is that just based on coincidences you have experienced or is
there something more that led you to that hypothesis?"

"It was coincidences that got my attention, but it was a nagging
feeling that kept bothering me." Jung explained.

Pauli knew exactly what Jung meant. "You feel like there must be a
scientific explanation, but you have no logical foothold to start from." He
said, describing the way he felt about the Pauli Effect.

"It wasn't until I stumbled on the law of attraction that made me
realize synchronicities might be something that can be studied." Jung's ideas
were still immature. "I am trying to conceptualize what it all is and how it
might be connected. So far, I have defined three separate things that could be
like different branches of the same tree. The tree is a symbol of our
consciousness and how it interacts with the world, or perhaps more

accurately, the tree is human-reality combined with our full conscious-unconscious minds."

"The tree is a conscious-human-reality?" Pauli said looking for clarification.

"It's also an explanation of how our consciousness may interact with reality." Jung had never shared these ideas and wasn't exactly sure how to explain it.

"What are the branches then?" Pauli asked, hoping they would add context to whatever tree he was attempting to describe.

"Synchronicities, the collective unconscious and the law of attraction." Jung said.

"I remember synchronicities but what are the other two?" Pauli asked curiously.

"The collective unconscious is the idea of an unconscious mind that does not belong to an individual but rather to an entire culture or society, or perhaps even the entire world."

"That seems nonsensical. Where would this mind exist? How would individuals connect with it?" Pauli challenged.

"No, no, it's not some external consciousness, though I do have some ideas for something along those lines that we might discuss another time. The collective unconscious is something internal to each individual. Learned or otherwise inherited like any other genetic trait or instinct." Jung explained.

"Nothing mystical about that." Pauli said casually.

"Not at all, though we do not fully understand evolution yet, seeing as how there seem to be periods in history where rapid evolutions took place as if there was some sort of choice to turn genes on and off, but I digress. [FACT 56] You are a good example of this collective unconscious."

"How so?"

"Once we decipher your dream of the dragonfly and your teacher, you will see that the symbols you are being shown carry a specific meaning. Why do you think I want to review my notes on those symbols?" Jung asked.

Pauli remained silent so Jung continued.

"Each symbol represents the same thing to you as it is does to other people."

"I'm not sure I follow. Are you saying if two people dream of the same thing, the dream represents the same meaning to both people?"

"Exactly," said Jung, "and allow me to remind you that the images you see in your dreams come from your own unconscious mind. Symbols are the language of the psyche just like how you described mathematics as the language of reality."

"That does seem strange, how did I learn this secret language? And what did you mean when you said the dreamer is shown the symbol? That's an odd way to say that, considering it is the dreamer who is showing the dream to himself."

"Ah, therein lies the conscious versus the unconscious mind. You and your physicist kin limit the definition of consciousness to our own everyday experiences but that only addresses the conscious mind. Your dreams were expressions from a part of your mind you've never acknowledged."

"A part of my mind…" Pauli grumbled as he trailed off. He never thought to ask where dreams come from but, until just now, he never understood the imagery in dreams is symbolic. "Alright, I see your point. You're saying the dreamer is shown the dream from some deep recess of his own mind. How do you define this hidden mind?"

"The idea of consciousness can be divided into conscious and unconscious experiences. The conscious mind creates a self-image of who we think we are, based on our virtues and cultural values but it blocks out any part of you that it doesn't like. On the other hand, your unconscious mind sees your whole self." Jung explained.

"Wait, we have a self-image that is not accurate?" Pauli asked.

"That is the ego."

"*That* is the ego?" Pauli asked. "I have never heard it explained like that."

"I prefer the term persona. The ego was Freud's crude explanation which lacked substance, so I developed the idea of archetypes. There are many archetypes but for now let's just talk about three; the persona, the self, and the shadow."

"Archetypes meaning what?" Pauli questioned.

"I don't want to distract you with all that now. Just think of the persona and the self as different ways your mind understands who you are."

"Okay," Pauli said. "The persona and the self are different self-images."

"Kind of. They are different *processes* the mind uses to understand who you are." Jung said, but before he could continue Pauli dove into an analogy.

"Like how computers run on code, but most people don't code, they use applications. Archetypal symbols are like the application, so people don't have to code, or in this case acknowledge their emotions." Pauli suggested to Jung who looked absolutely flabbergasted.

"Honestly, I don't understand anything about computers, so I have no idea what you are talking about," Jung said. "The archetype 'self' is the whole self as it truly exists in human reality. It encompasses the good and the bad of who you are and what you have done. You cannot hide anything from the self. The 'persona' is the ideal version of ourselves that we believe we are showing to the external world."

"Like a mask we wear." Pauli suggested.

"The thing is the mask isn't always accurate. You see, people unknowingly pretend to be more decent human beings than they are. The unconscious mind sees the 'self' and the 'persona' at the same time and when the two don't match the 'shadow' is created."

"When the brain's two self-images don't match, we get a shadow." Pauli confirmed. "What kind of shadow?"

"Think of it like a literal shadow. Imagine you are looking at someone - or better yet, you are looking at yourself in the mirror."

"Okay."

"Well in the mirror you can see your 'self' but you are wearing the mask of your persona."

"The persona is blocking the true self." Pauli understood.

"Yes, but only partially. It's casting a shadow on the parts of the self that don't match the persona." Jung lectured.

Does a mask really cast a shadow, Pauli thought as he realized he did not like Jung's analogy. Pauli pictured what Jung was saying in his own way.

Instead of imagining himself looking in a mirror, Pauli pictured an avatar of himself standing across an empty room. This was the whole 'self' which included his good and bad parts. In front of the Pauli-avatar was a giant shield, shaped like Pauli's face which looked like a clear masquerade mask. This is the persona that I show the world, he figured.

Pauli looked down and saw he was standing on a label that said, 'Conscious Perspective.' As he looked at himself from this angle, through the masquerade mask, he saw the best version of himself. He saw another label on the ground that said, 'Unconscious Perspective.' He walked over and stood on it. From this angle he was looking directly at his self, not through the persona's mask. He saw his good and bad parts. When he stepped back to the Conscious Perspective, all the bad parts were filtered out again. This must be what the conscious mind sees, he thought. When it looks at the self, it sees a filtered version. Ever the scientist, Pauli looked closer to see what was happening.

For us to see things, photons reflect off objects. When they bounce, they take on the frequency of the thing they hit, giving each photon a specific color. The color-tuned photons then hit our eyes, where our brains create pictures in our minds, Pauli thought, fully digesting this analogy.

When the persona is between me and the self, some of the photons are being filtered out by the persona before they reach me. It scatters the photons like one of those invisibility shields, so I don't notice that anything is being blocked at all. [FACT 57]

"The shadow is invisible to the conscious mind." Pauli said definitively.

Jung was a little confused by this abbreviated response but said, "yes, I think you've got it."

"But what is actually being filtered out?" Pauli asked.

"The aspects of yourself that do not match your values or your conscious self-image. Your conscious mind pretends there is no shadow while your unconscious mind, not only knows there is a shadow, but also sees what the shadow is hiding." Jung explained. [FACT 58]

"Okay," Pauli said. "We have the shadow which blocks the full truth of the self, we have the full self and we have the persona which is what other people see of us."

"Not quite," Jung corrected, "this is an important distinction. The persona is not how others see you. It is how you *think* others see you."

"Oh." With that Pauli understood the persona is not the version of him that actually exists in human reality, instead it is the self-image he believes exists in human-reality but others might see his true self. "This is all symbolic. What does any of this actually look like in reality?" He asked abruptly.

"The unconscious mind sees what the shadow is hiding and tries to inform the conscious mind through dreams and feelings."

"Now I see," Pauli said. "You've been talking about dreams this whole time. Our unconscious mind chooses the dreams our conscious mind sees. It is still our mind showing it to ourselves."

"That's right, but dreams aren't the only way the conscious mind sees the self. Sometimes other people tell you what your shadow is hiding. When that happens, your ego might cause you to act 'out of character.'" Jung explained.

"Act?" Pauli interrupted. "How could some hidden part of the mind cause someone to act? Wouldn't that person think something very strange was happening?"

"The ego covers its tracks! Herr Pauli, it was only this morning you sat in my office and we discussed how you, yourself, lashed out at your wife before she left you. You told me after you yelled at her, you 'didn't know where that came from.' Do you recall? It came from your unconscious mind!" Dr. Jung knew Pauli had been close to a breakthrough in that morning's session. He didn't want to push ahead and force a realization, but now it was too late.

Pauli was stunned. He thought he was better for going to counseling, but this was an eye-opening realization. "But I wasn't lying to myself about anything! ...was I?" He thought aloud.

"One of your wife's complaints was that you worked too much and spent too little time with her, correct?" Jung asked.

Pauli held up his hand as if gesturing for Jung to stop. "I didn't lie to her or myself about any of that, I knew I worked a lot and not only was it justified but I told her it would be that way before we married. She knew what she was getting into!" Said Pauli.

"Yes, my friend you did, but what you just said is the perfect example of the conscious mind covering its tracks with a logical explanation." Jung paused, allowing a moment for the harsh reality to set in. "What you just said, that logical explanation, justified your behavior and allowed you to keep making the same choice day in and day out, allowing the problem to grow. Here is your realization my friend. What your shadow is blocking is the fact that you felt guilty about neglecting your wife all along and your unconscious mind knew it."

After a long sip of coffee followed by a few moments of introspection, Pauli responded. "I did feel guilty about working long hours. I remember the first time I noticed her loneliness. I felt bad but I thought about the breakthrough I was working towards in the lab and that was the first time I told myself what I just told you. I feel like I was in a no-win situation. Had I given in and spent more time with my wife then I would have felt guilty that I wasn't spending more time working." Pauli admitted for the first time.

"Perhaps you and your wife were not as compatible as you first thought." Jung offered, but Pauli was making peace with it in his own way.

"And the shadow suppressed my true feelings, pushing them down into my subconscious mind in order to protect my persona's self-image." Pauli started to sound more like the tired man who first stumbled into Jung's office. This medicine did not taste good, but he knew he needed to take it so he pushed on. "I always thought of myself as a good husband, that was my persona. I neglected my wife which was something that a good husband wouldn't do. The self and the persona did not match and that created the shadow. My ego then protected the belief I was a good husband with a logical explanation."

"Which caused the shadow to grow." Jung gently added.

Pauli was defeated but wanted to get through it, so he spoke plainly about how all this had actually played out.

"When my wife tried to talk to me about how I spent my time, she was making me realize that I felt guilty but I was too busy pretending I had done nothing wrong. I hadn't been as good of a husband as my conscious mind had me believe. The self-images clashed and instead of acknowledging the guilt I felt I yelled at my wife because she pointed it out," Pauli realized as his analytical voice turned to a softer tone. "All she did was ask me to spend more time with her and I yelled at her for it." He said morbidly with a long sigh before taking a few deep breaths.

Dr. Jung realized it was inappropriate to bring all this up in such a public setting but at least the bustling city provided a rumble of sound to veil their conversation. He tried to think of something to say to break the silence but it was Pauli who spoke first.

"I suppose I would have asked for a divorce too!" He said trying to laugh.

"When the truth is first exposed it can seem quite ugly. The important thing is that you acknowledge just how ugly it is so you can correct it." Jung said optimistically. "And now you see what I meant when I said we all have a self-image that is a better version of who we really are." Jung said.

"So, we each have a mind; half conscious, half unconscious." Pauli surmised. "The conscious part is all our experiences and thoughts including how we think the world views us. The unconscious mind knows all that but also accounts for our emotions and actions; our full self."

"Yes." Jung said quietly, still gauging Pauli's mood.

"Well, don't look so surprised," Pauli said as positively as he could, "you're learning physics, so I might as well learn your specialty while I have the chance!"

"But this just covers the individual." Jung said, trying to intrigue Pauli. "What I want to share with you is the collective unconscious mind. I have studied many individuals and found that the same symbols represent the same things."

"That does seem strange," said Pauli. "Why would different people have symbols that mean the same thing? But certainly, different cultures have different symbols?"

"They do, but there seem to be some symbols shared across all humanity. I believe each individual simply inherits knowledge of these symbols through the same genetic inheritance they obtain instincts." Jung theorized. [FACT 59]

"That would make sense." Pauli agreed.

"The interesting part is that we can actually contribute to the pool of symbols in the collective unconscious!" Jung said, now speaking quickly, excited to be sharing his theories.

"Contribute how? Are you suggesting some great pool of ethereal knowledge we all draw from or somehow tap into?"

"Perhaps, yes, but I cannot say anything scientific about that at the moment, so I will not engage that discussion just yet. Instead, I offer the following explanation. It is the obvious explanation that you will find quite reasonable."

"Very well."

"Let's say someone writes a popular play that has unique and specific imagery and symbols. This play becomes wildly popular causing some of its symbols or phrases to be added to society's everyday vernacular."

"You can have your cake and eat it too." Pauli blurted out.

"Good example," Jung said. "If you dream of having cake but you are sad because you can't eat it then the writer of that play successfully added to your unconscious and if others did too then it was added to the collective. The thing about the unconscious mind is, it remembers everything."

"How is that?"

"The unconscious is our intrinsic memory." Jung explained. "Intrinsic memories are formed based on feelings and emotions, not words."

"So, our normal conscious memory is words, like a dialogue or reading a transcript of the scene you are remembering?" Pauli guessed.

"Exactly. That is explicit memory. Do you have any memories from before you could talk?" Jung challenged.

"No," Pauli shrugged.

"That's because what you think of as your memories are only your explicit memories based on words." Jung explained. "What I'm talking about is your intrinsic memory which is how we draw from the collective unconscious. It's based on the feelings associated with external stimuli."

"I see," Pauli thought. "And could these understandings be physical things? I mean inside your brain. It feels like the collective unconscious you are describing is like a physical thing."

"Probably not, but that is the great mystery with consciousness. There are brain patterns we can recognize, but there is no physical thing we can point to and say this is consciousness." Jung explained. "Therefore, the collective unconscious is maintained by the individual minds of all people. Even the most recluse humans are still part of a species that is inherently social."

"And this social aspect is what allows for a collective," Pauli understood.

"Exactly," Jung explained, "people use the same symbols in stories and traditions creating the cycle of internal and external understandings. But now let me tell you about the collective shadow!"

FACTS

56.   What you may have learned in 10th grade biology about your genetic inheritance is wrong. It is possible for you to consciously change how your genes are expressed. It is also possible for those changes to be passed on to your kids. [Ref 47]

No, you can't change genes just by thinking about them changing. You have to adopt behavioral changes in your life that create chemical changes in your body which affect certain genes which have plasticity (the ability to change). This is called epigenetics.

As the genome randomly mutates slowly over many generations it is also likely that there are more rapid periods of genome adaptations triggered by changes in the external world.   [Ref 47, 48]

Do not take this realization lightly. You can turn on and off genes to become the person you want to be. Remember, life is all about taking action.

57.   Invisibility shields are a real technology we have today. The design is quite simple. When an observer looks through the shield the photons are scattered away from the observer. They are either dispersed to either side or focused immediately in front of the shield. An observer standing further back would not see the hidden object (if the shield is in the correct place, relative to the object being hidden.)

You have likely seen and even touched this technology yourself. Have you ever seen a poster where the picture changes when you move from left to right? That is the same thing; but for an invisibility shield, the screen is clear and slightly curved to create the lensing effect. [Ref 49]

58.   The lies your *persona* is telling you are not restricted to values or cultural ideals. All the ways you self-identify contribute to your *persona*. For example, if you feel you are an open-minded person then you are probably very tolerant to people who are different than you. However, if you feel that a group has wronged you in the past then you are, most likely, not open-minded towards people in that group, which inherently clashes with the idea that you are open-minded.

In this scenario, your *persona* is an open-minded person but your *self* is a selectively open-minded person who has closed off a certain group of people. [Ref 27]

Your logical argument might be something like 'I remember how that group treated me' or 'I see how that group treats someone from another group; therefore I know how closed-minded they are.' Do you have a *shadow* your ego is hiding from you? Looking at you Democrats and Republicans. [Ref 55]

59.   It is not accurate to say we inherit our instincts genetically. It is more complicated than that. What we inherit genetically, is the ability to learn our instincts, or in this case, the symbols our unconscious mind uses. The genome allows for instincts to be learned in a way that is specific to the individual which does not change the individual's DNA. [Ref 50]

Chapter 20

History Erased

"The collective shadow." Jung said, still sitting with Professor Pauli at the café in the heart of Zurich. "The collective shadow is the same shadow concept as the individual however instead of the individual's own actions creating it, it is created by the actions of other people within the same group."

"Other people's actions that make you feel guilty somehow. How does that work?" Pauli said.

"Our persona automatically assigns various groups to its own self portrait. This stems from a tribal mentality that was necessary for our ancestors to survive in the wilds and early civilizations."

"Us versus them."

"Exactly. If you self-identify as a Christian, then you belong to that group. When you belong to a group you gain a personal responsibility for that group's actions; good and bad."

"Can you give an example?" Asked Pauli.

"I'm a Christian but if some other Christians do something awful, like what they did during the Crusades, non-Christians might look at me and think I support the awful things they did."

"That's a messy example." Pauli challenged. "That was almost a millennia ago. So much has changed since then."

"That's true," Jung relented. "Plus, Christianity is not defined by the Crusades. There is a better example over in America that's still relevant to this day. I predict the Americans will be plagued by racial conflicts that flair up every few years until their collective shadow is properly addressed."

"I like this example." Pauli said rubbing his hands together. "Real world events to explain the intangible subconscious mind."

"In America they kept humans as slaves based on their race. This was such a big issue they fought the Civil War over it. The Union was the side fighting to end slavery. They won the war but instead of celebrating their heroes, the Americans kept all the symbols from the losing side, the Confederates."

"So what?" Pauli challenged. "They're just symbols. How does a symbol contribute to this collective shadow you are describing?"

"Symbols are more powerful than you realize. Nowadays there is no one left alive who fought in the Civil War or kept slaves but there are lots of people who still identify with Confederate symbols. These people did not do the bad things, but their group did."

"The public 'persona' and the whole 'self.'" Pauli mumbled aloud, reminding himself of the archetypes associated with the 'shadow.'

"Let's talk about two theoretical people, both born at least 30 years after the Civil War ended. The first person is a descendent of Confederates while the other is a descendent of slaves.

The first person grew up surrounded by symbols of the Confederacy. They would naturally understand the Confederate flag represents their heritage, because it represents a group they belong to. The thing is, they were not there to see what the flag fully represents."

"Ah, there is the shadow." Pauli said, noticing the partial self-identity. "This modern confederate could genuinely be a good person." Pauli realized.

"Yes exactly. Now to the second person. If the descendent of slaves saw the Confederate flag being proudly displayed by the first person, they wouldn't see some proud heritage. All they would see is a symbol of the group who enslaved their ancestors, and by extension, may even want to enslave them."

"So, a storm is brewing. One group of people identify, innocently, with a symbol that represents evil to another group. All the while both groups are filled with reasonable and otherwise good-hearted people." Pauli summarized.

"Neither individual has done anything wrong, yet they are *designed* to hate each other." Jung finished.

"Yes, I can see how this will lead to issues in the future but I'm not sure I understand the idea of the collective shadow." Pauli said.

"The Confederate people might acknowledge the fact that slavery was wrong, but they don't see their group's role in it. They allowed their egos to protect their positive self-image. Protecting the self-image required them to bury the truth of what people in their group did." Jung explained.

"That makes sense."

"Allow me to make a distinction." Jung continued. "We can split all the Confederates into two categories. One type does not know their group did anything wrong because they never saw or learned the truth. The other type of individual knows about the evil but still chooses to identify with the group for one reason or another. Both groups self-identify with the symbols of the Confederacy but only the ones who know about the evil deeds feel the effects of the shadow."

"You are saying there are people who legitimately don't know the Confederates' whole ideology is based in racism? How could they not know? It seems so obvious."

"When you are experiencing the emotions for yourself it is never obvious. How did you suppress your shadow when you neglected your wife?" Jung challenged.

Pauli grumbled and sipped his coffee before finally giving an audible sigh. "I used a logical argument to ignore my feelings." He said bleakly before asking, "But wait, if these Confederate-lovers are ignoring their feelings with logic, what logical argument do they use to justify slavery?"

"That's just it, they don't address slavery at all. They say the war was fought over States' rights for economic freedom, not slavery. On the surface this is a legitimate argument but here's the thing; they were fighting for the economic freedom to do what?"

"Keep slavery." Pauli reasoned without much effort. [FACT 60]

"Exactly." Jung said.

"And you said the shadow only affects the ones who know they did bad things?" Pauli prodded for more explanation.

"Well, it's more complicated than that," said Jung, who felt like Pauli during a physics lecture, flexing his knowledge over the other man's ignorance. "Remember the shadow is only created when an individual's self-identity (persona) clashes with their actions (self). In this case it is the actions of the group they belong to."

"I see now," Pauli said. "The people who don't realize their group did anything wrong have no shadow because the persona and the self match."

"I will add that people only fall in this category for a short time, basically this only applies to children. Once they hear the slightest suggestion that their group represents something bad, the persona and the self begin to offset."

"At which point they fall into the category of Confederates who do know their group did bad things, so does that make them all racist?" Pauli asked, not sure if he was following Jung's thought.

"What is it to truly know something?" Jung challenged.

Pauli was confused. "The scientific method proves truths so we can gain an understanding." He brainstormed aloud.

"Is an understanding the same thing as knowing something?" Asked Jung. "At what point would the confederate *know* their group did something wrong?"

"I see your point. This subject is not an all-or-nothing situation, there is a middle ground where they have an inkling that something might not be right but they don't necessarily know." Pauli understood.

"In your line of work everything may be black or white, but my work begins and ends in gray matters where all human experiences live. In other words, no, just because someone identifies with confederate symbols, that does not automatically make them racist." Jung explained. "We can divide this group into two more groups."

"This is getting complicated." Pauli sighed.

"Out of all the Confederate supporters who are aware of the evil deeds; one group is overcome by their pride in heritage. They may feel there is something bad hiding in their group's identity, but they allow the logical arguments to suppress those feelings. The other group, which is a small minority, actually support the evil deeds."

"The last group is easiest to understand," Pauli said bluntly, "those people are racist."

"Yes, but I believe that is a small percentage." Jung explained. "Think about where these people's persona and self come from. What consciously unifies this group?"

"I'm not sure." Pauli said.

"It is not racism that unites modern confederates but the symbols themselves. Most have never done or seen anyone in their group do anything wrong, so naturally their self-identity, their persona, is someone who is not racist because they are not racist. Then someone comes along and tells them the group they belonged to is racist."

"Deep down they feel like it might be true," Pauli understood, "which is the self clashing with their persona."

"Because of the values they hold," added Jung.

"But when someone tells them their group is racist why does that person accept what they said? I mean, why not just ignore the people accusing them of being racist?"

"At first that is exactly what they do," Jung said, "but as you can see, the symbol actually is racist so they cannot escape it. The conscious mind repeats the logical lies but the subconscious mind begins to realize the hidden truth which starts to eat away at them."

"And the shadow arises to protect their egos so they can keep thinking of themselves as virtuous. It's a gradual process then." Pauli realized. He thought for a minute before saying, "Actually, when the shadow makes them upset, in a way, it means their values align with the person calling them out."

"Only when they are genuinely upset." Jung explained. "The true racist might feign anger so they can blend in. The Confederate person who genuinely does not like being labeled a racist was taught to identify with a group but they do not want to be associated with racism. That is why they get upset when someone points out the clear connection the confederate flag has to it."

"And if the person is racist there is no shadow at all." Pauli figured.

"That's right. The symbol is racist and the person is racist, so the persona and self match."

"Alright, I think I understand." Pauli surmised, "The collective shadow is created when some people in a group know the full truth while others have suppressed the truth."

Jung paused. That's not right at all, he thought. "No, no, no! You're over thinking it." Jung felt a jolt of satisfaction when he got to tell Pauli he was wrong just as Pauli had been doing to him. "The collective shadow was created when the former Confederates put up statues of their leaders everywhere and effectively erased the bad parts of their group's history."

"They erased history?"

"They only fooled themselves. The collective shadow rose when they taught their children to understand the Confederate flag was not a symbol of racism and instead something to have pride in." [FACT 61]

"The next generation never saw the evil and they were taught it never existed in the first place." Pauli understood, somewhat in awe of how clouded the situation seems to someone who is experiencing it, compared to how easily he understood it all from the outside looking in, emotionally uninvolved.

"In a way, it was the winners who failed. They could have outlawed the Confederate symbols and created new symbols to represent the reunification of one nation." Jung pontificated. [FACT 62]

"At the very least they could have used their existing symbols to celebrate reunification." Pauli figured. "I have witnessed the rebel spirit of Americans before. They love to show it off but it is the rebel spirit of the revolution, not the Civil War, that should be celebrated."

"To finish this thought, if the righteous victors had abolished the symbols and forced the Confederates to acknowledge their evil past, it would have allowed the next generation to move on." [FACT 63]

"Instead, they passed on a generational trauma and the collective shadow will grow." Pauli thought aloud.

"Precisely, and the decedents of the slaves will forget that over 350,000 white Americans died so they could be free. [FACT 64] Whenever they do decide to address this shadow, it will only take one generation to heal from the whole thing."

"How quickly things can change." Pauli remarked.

FACTS

60.   Slavery was the primary contributing reason the American Civil War was fought. [Ref 51, 52]

Did every Confederate soldier join the army to keep slavery? No. Many allowed themselves to be swayed by the political voices of the time who called for economic freedom to maintain their culture and their 'peculiar institution.' Throughout history and today, any reasonable person can ask themselves the same question Jung asked Pauli in the narrative; states wanted economic freedom to do what?

You may be thinking, it's so obvious, but hindsight is 20/20. What are your political leaders telling you about the issues of today that will have a different interpretation years from now? Remember, it is not what our leaders say that matters, it is what they do. After all, the road to hell is paved with good intentions.

If people on the other side of the aisle are continuously disputing an issue you think you fully understand, then perhaps you should gain a better understanding of how the issues are being presented to you. If a 'truth' is so obvious to you, why is it not to them? *Because they are bad people*, is what we romantically allow ourselves to think. If that's what you think about your fellow Americans then you may be too emotional about the subject to see the truth. The truth is *those* people think their position is the obvious one. They don't understand why you don't see it their way.

61.   People who self-identify with the confederate flag have erased part of their history. Need proof? Just look at the flag. You probably think the red flag with the diagonal blue stripes was the flag representing the 'country' of

the Confederate States of America. Almost. It started as a battle standard; a flag used in a few battles that only represented one confederate unit. It was incorporated as one corner of the Confederate Flag but the 'Stars and Bars' were never the whole flag. In one design the rest of the flag was just white, which is way worse, but at least it was honest. The Confederacy was a white country for white people. When they cut off the white part of the flag, they were cutting out that fact from history.

A group called The Daughters of the Confederacy thought it was a great design to represent the self-image they were fabricating. [Ref 53] The Daughters of the Confederacy then (easily) convinced local municipalities throughout the South to 'celebrate their heritage.' Instead of acknowledging the fact that slavery was wrong, they allowed their egos to protect their positive self-image by burying the truth of what people in their group did. [Ref 54]

If you agree with 100% of your political party's ideas then you are contributing to the negative consequences of groupthink that George Orwell tried to warn us about. No one who has their own thoughts agrees with 100% of their political party's views. Don't let your individual voice be washed away with the masses. If you have your own thoughts, then you have power.

62.   In America, our First Amendment rights protect symbols from being outlawed. This is a good thing. It ensures that our government, and those who are elected to power at any given time, cannot censor things they don't like. Doing that would create a much worse collective shadow for us all.

Instead of censoring symbols we should call them out publicly for what they truly are. I'll go first: Confederate symbols represent losers who lost at being racist.

63.   Where are the statues of Hitler? After WWII, instead of imagery celebrating the 'heritage' of the third Reich, the Germans built memorials to the murdered Jews. They faced the ugly truth of what other people in their group did. They addressed it and are, today, a healthy society. They learned firsthand that each individual must not allow groupthink to poison society.

It is ironic that modern confederates claim the removal of their statues are erasing history when the truth is those statues are a cover story. Had the Germans 'celebrated' their pride in heritage by putting up statues of Hitler they would have been hiding the real history of what happened just as many Americans are still fighting to keep the truth of the matter hidden behind proud statues of heroic looking men. It's time to face the truth of history and leave the Confederacy behind.

64.   The Civil War produced 642,427 total *Union* casualties: [Ref 56]
        -110,100 people were directly killed in combat
        -224,580 people died of diseases because of their involvement in the
         war
        -30,192 people died while prisoners of war
        -Another 275,174 people were wounded in action

Over 367,000 Union soldiers, mostly white, died fighting to free the salves. Did this blood pay the debt that slavery created? Perhaps, yes; perhaps no. Some people are calling for reparations citing damages in their current lives. Perhaps a more direct cause of these damages isn't slavery itself but the laws established after the Civil War designed to keep the black population oppressed.

If those laws had not been passed, and upheld, then black folks would have become successful in society as they did in places like Black Wall Street. In Tulsa, Oklahoma free black people lived in peace and prosperity. The bombing of black Wall Street in 1921 is just one example of actions that are more direct causes for reparations.

The idea of paying reparations is not as simple as cutting checks and mailing them out. Who would pay reparations? The government that black people pay taxes to? That would mean they are paying themselves. Should we force the descendants of those who held slaves to pay? How could we prove the pedigree of all the perpetrators? For that matter, how can we prove who is a descendant of someone who was directly victimized? Are we only paying reparations to people who can prove they descend from former slaves? What about people who are descendants of both slaves and slave owners, do they pay themselves reparations? Should we pay people who just stepped off a plane from Africa simply because they are black, even though they have no ties to the Black-American experience?

If the damages are that black people are born into poverty at a higher rate than whites [Ref 57] then the reparations should be paid in the form of assistance that will allow black folks to pull themselves out of poverty, just as they did on Black Wall Street. Measuring success that way is dangerous, we cannot force outcomes. We can, and should, make it easier for anyone born into poverty to work themselves into prosperity. Is that not the American dream?

Chapter 21

The Burden of Proof

"This collective shadow will linger until they acknowledge confederate symbols cannot be separated from racism." Jung culminated. "But I digress, I want to talk to you about the last branch on that tree I was telling you about."

"There is another branch?" Pauli asked. "Remind me what the branches are again."

"Synchronicity, the collective unconscious, and the law of attraction."

"Why are these concepts different branches on the same tree? What connects them?" Pauli asked.

"That's just it, I believe the explanation for one will be related to the explanations for the others because of their similarities." Jung felt like he was on a roll, so he let his true thoughts come out. "I believe consciousness may be the trunk of the tree and the more I talk to you the more I believe physics might be the roots that hold the whole thing up."

Pauli thought for a moment before saying, "I'm not sure I understand your tree but I do appreciate physics being the roots. To me, physics is synonymous with reality itself, therefore physics is the root of all existence. Now, what is the Law of Attraction?" He asked.

"When I first read about it, I thought it sounded like someone selling snake oil. It all sounds too good to be true. Multiple authors make different claims that are all, supposedly, based on scientific principles. As a physicist, have you ever come across the Law of Attraction anywhere?"

"No," Pauli said shaking his head. "I've never heard of this law; is it from some other discipline?" Pauli asked.

"Well, I'm not sure," said Jung. "They seem to think it is an explanation of how the universe works which is how you described physics. The law of attraction postulates that our consciousness somehow creates reality. 'Thoughts are things' they say."

Pauli scoffed, "Sounds like new-thought nonsense to me. I cannot see how our thoughts affect reality."

"Well, I could break that up into two categories. There is a very obvious way your thoughts can, and do, affect reality and then there is a theory of something more. So let me ask you this; can I use my mind to pick up this cup of coffee?" Jung asked.

"Of course not, but you could use your hand."

"But my hand didn't decide to pick up the coffee. My mind told my hand to pick it up."

"But to say you picked it up with your mind, implies you used *only* your mind." Pauli argued.

"Yes, I agree, but take the word games out of it. Did my mind use the tools it had to pick up this coffee?"

"Well, if you say it like that, then yes." Pauli said.

"That is the obvious way our minds can shape our reality; directly. And isn't this the perfect way to frame the mystery of consciousness." Jung added.

"How is that?"

"Well I consciously decided to pick up the coffee. What if I decided to throw it into the street rather than drink it?"

"What nonsense are you talking about?" Pauli chuckled.

"If I threw that coffee in the street the broken mug might flatten someone's tire, drastically changing their day, but I didn't so it won't. I'm just making the point that in our shared objective reality things happen, and other things don't happen because of our conscious choices." Jung explained before asking, "how and why do I make the decisions I do?"

"You have free will, don't you?" Pauli said, unamused. "You thought about it in your mind and decided the best thing for you would be to drink your coffee rather than throw it because it will nourish your body. Where is the mystery there?"

"All animals have brains, do all animals have consciousness as well?"

"Probably not," Pauli said casually.

"Then where do we draw the line?" Jung challenged. "Is it in the size of the brain?"

"Perhaps." Pauli said.

"Wrong," Jung replied calmly. "Birds have notoriously small brains, but some species have been proven to exhibit behavior that requires some form of conscious thought." [FACT 65]

"That is actually surprising." Pauli admitted.

"Furthermore, just like how gravity is missing a quantum particle, there is no physical thing we can point to and say this is what causes, creates, or allows for consciousness." Jung explained.

"Ok," Pauli admitted, "I am beginning to see the mystery of consciousness."

"Now that we understand our consciousness can influence reality in this most direct way, it begs the question - what are the limits of what our minds can achieve?" Jung thought aloud.

"What do you mean?"

"I'm not sure about consciousness literally creating our reality but at the very least the Law of Attraction is functionally accurate." Jung declared.

"How can you claim such a thing?" Pauli challenged.

"I believe many people think it works like magic, so they forget the basics."

"What basics are those?" Pauli asked impatiently.

"It's like a process with steps. First, you decide what you want. Second, you hold in your mind, the idea of yourself having the thing or living the life you want as if you have already achieved it. Third, you stay in Step Two until you genuinely feel grateful for having already gotten what you want."

"Even though you haven't gotten it yet?" Pauli interrupted. "Sounds like a bunch of mind games to me."

"That's where some people stop, then they claim it is all nonsense. The fourth step is to receive divine inspiration." Jung explained.

"Okay, now you've lost me. Step four is to hear God speak to you?" Pauli said sarcastically.

"Perhaps divine and inspiration are the wrong words. The fourth step is to figure out what you need to do, in reality, to get what you want." Jung explained.

"That is not at all what you just said!" Pauli claimed. "How does divine inspiration come to mean, make a plan of action?"

"Our intuition," Jung said confidently.

"Intuition?" Pauli sounded skeptical.

"That's right, I called it divine but really what I'm describing is just a conscious thought that pops into your mind from nowhere." Jung paused his lecture to let the point sink in before asking, "have you ever had an answer pop into your mind from nowhere? It could have been a eureka moment or just a nagging feeling."

"When I contemplate physics, I often have solutions simply appear in my mind; seemingly from nowhere." Pauli relented.

"Back to the law of attraction." Jung steered the conversation. "The fourth step is to listen to that intuitive solution when it pops into your mind. It lays out a path for you to get what you want. Some people remember this step but stop here. The final step is the most important."

"Yes?"

"Act. The person must act upon their intuitive thoughts, and they must do something to achieve their goals."

"That was anticlimactic." Pauli joked.

"For every single thing that has ever happened in our reality, has anything ever happened for nothing?"

"Well, no," Pauli said, not sure if he understood. "Energy cannot be created or destroyed so for something to happen something else must cause it."

"Well, the same is true for the law of attraction. What you want comes at a price and the most common price is the work it takes to get whatever it is!" Jung explained. "And so I ask again, what are the limits of what the mind can achieve with this direct method of the law of attraction? If

The Burden of Proof

I had a business that I wanted to build into an empire then I would need to start by defining that goal, consciously, in my mind. Then, if I learn enough and work hard enough, I can start to actually create that empire. Is this not how Carnegie and Rockefeller have amassed their fortunes?"

"I suppose so, but you seem to be focusing on the fact that they made a goal when in reality those men were smart, lucky and willing to work very long hours." Pauli argued.

"Of course there was much more to it, but every step of the way, every bump in the road, every time something happened that made them question their goal; they chose to continue on. At any point in time, they could have given up and their dream would have immediately dissolved, never to be realized in human-reality. So even in this direct way it is a thought that is the very foundation of what is created in our reality."

"I can agree they had to continue focusing on their goals and using their thoughts to navigate how they would achieve them, but this is such a reductionist view." Pauli argued, sounding a little irritated. "There is nothing special about this process and it requires no special designation as some sort of 'Law' of the universe."

"But we are scientists," Jung challenged. "Isn't our duty to describe the unknown and assign definitions and labels? You said yourself, the entire field of physics is just a description of the universe and how it works. My profession is a description of the human experience within your universe. If this is the process one can use to achieve their goals, then it should be defined."

"I see your point." Pauli said more calmly, "but this seems like you are just describing the steps someone took to achieve a goal, I don't see how this is some scientific law."

"What I described is measurable and the results can be repeated."

"No!" Pauli exclaimed, letting out some frustration by pounding his fist on the table, startling another patron of the café who almost spilled her coffee. "The same results would not be reproduced. For one successful business turned conglomerate there are a hundred, or perhaps a thousand that have failed."

"Yes, but in those failed cases, did the person driving the vision give up rather than double down on their goal? I suppose that is a problem

235

though." Jung said with a sigh before Pauli could retort. "Not everyone who wants to be a business mogul will succeed so how do we define the exact moment when an individual gives up."

"There are too many variables for a legitimate experiment; you would never be able to test this hypothesis!" Pauli said.

"And yet," Jung said calmly, "if all successful business moguls followed a method like this, then there is value in defining it."

"How can you claim they all use this method?" Pauli challenged.

"I can't." Jung said plainly. "However, I have studied some and they all seem to have defined their goal then worked relentlessly to achieve it. They each understood they could change the world in a very real sense. Once you realize you have hands, it is easy to lift the mug." [FACT 66]

"Do you think that is all that is being referred to in those books about the law of attraction?" Pauli challenged.

"No," said Jung honestly. "They speculate some unseen forces. They describe some external consciousness, some aspect of ourselves that also influences the world we live in. Actually, I believe it's more than that. It seems to me the only way to explain it requires all of reality to be the manifestation of some great consciousness that we can all influence. As I read about those concepts I was torn. They sound like fantasy, yet they resonated with me. Something about them feels right, especially when considering explanations for my idea of synchronicity."

"But how would any of that work?" Pauli challenged.

"When I think about how they could be explained I always get stuck at two core questions. Where did our reality come from and what is consciousness?" Jung said.

"I finally understand your quest for this knowledge, misplaced as it might be." Pauli said.

"Misplaced?" Jung said, angered at Pauli's callus. "It was only last week you were describing the so-called Pauli Effect. You are so convinced it's a real phenomenon that you went on and on trying to convince me. Do you believe the Pauli Effect is real or not?"

"My belief is irrelevant if I cannot prove it in a way where the results can be reproduced!" Pauli blurted out.

Jung paused. "You make a valid point. For the science to be valid it must be testable. But in this case, we don't know where to begin to create a test so I feel I must suspend my disbelief and doubts. This way I can explore its implications before anything is proven. The problem is, once I have some elementary understanding, I will be like an island of knowledge waiting for others to come to the same conclusions."

"Perhaps I will build a bridge and accompany you on this island. I can suspend my need to prove the phenomenon so that I may one day understand the blasted Pauli Effect." Replied Pauli, trying to take a leap of faith so he could explore the unknown possibilities of reality.

"So, it's still happening?" Jung asked.

"Yes. It is." Pauli said rather quietly.

"If I was to categorize the Pauli Effect into either synchronicity or the law of attraction, I wouldn't know which to use. Perhaps it started as a synchronicity but now that you are aware of it, you expect it, which is a principle of the law of attraction." Jung said confidently.

"The burden of proof is too great." Pauli said abruptly, clearly still clinging to the scientific method, "how would you test for any of this?"

"So that's it then?" Jung challenged calmly, "We cannot figure out how to test it objectively, therefore it must not exist at all? Did gravity not exist before Newton defined it?"

Pauli sighed and finally let go of his rigid hold on the scientific method. After a moment he said, "do you know it happened to me again the other day?" He sounded tired of the whole thing.

"What was it this time?" Asked Jung.

"My colleagues enjoy tormenting me so much that at a social event they rigged up an old chandelier to fall to the ground as I entered the room. As it happened, I walked in the room and the person activated the rig only to realize that it malfunctioned and stayed fixed to the ceiling!" [FACT 67]

Jung couldn't contain his laughter. "Imagine that! It malfunctioned by not breaking!" He laughed so much that Pauli even began to laugh at the whole thing. "Even in jest you unknowingly break things. I have to tell you Herr Pauli, I would be thrilled if this sort of thing was happening to me, it would be a constant reminder of why I must explore what causes it!"

"Perhaps it is." Pauli mumbled thoughtfully. "Remind me again, what your definition for synchronicity is?"

"Two or more acausal stimuli, appearing in a way where the observer can see a clear connection. Like when two people go on vacation and happen to run into each other on the other side of the world. Logic would tell us they could have ended up in the same city without running into each other, or that they may have been in the same place but at different times, or even in the same place and time but didn't notice each other in the crowds. What are the odds they actually see each other?"

"Yes, but how often does something like that happen and how do you separate that from a random coincidence?" Pauli challenged.

"What is a coincidence? Is it not true that every time your curse has broken something it was just a mere coincidence?" Jung suggested.

"Alone, each one is mere coincidence. But there is some consistency with my experiences that seem to not be present with that example. For me, this phenomenon has been happening for years, in your example it is one time." Pauli explained.

"This is a perfect description for how I study everything in my field. I do not study masses of people, only individuals. When I saw the same dream symbolizing the same thing in many individuals, I pieced together my theory of the collective unconscious." Jung explained. "The parts are not separate from the whole."

"A fractal," Pauli said. "It's something that has the same shape as its parts." He took a moment to collect his thoughts. "Are you saying that every coincidence is a product of the same creation process?"

"I have spent many hours thinking about this, and yes. No matter how significant or insignificant each coincidence; either they are all random chance and the Pauli Effect is all in you and your colleagues' heads, or every single coincidence is created by something that is, at this time, known only to the universe." Jung said mystically.

"You know something? The branches of your tree could be related to quantum entanglement. Einstein hates it, he mocks it as 'spooky action at a distance' and he seems right to do so. If Bohr is correct and particles do, in fact, have some connection that can influence one another over vast spaces, then perhaps there is some excitation in those particle fields that actually connects these strange coincidences." Pauli thought aloud.

"You've lost me, what is this spooky action at a distance?" Jung inquired.

"Entanglement. But I'm afraid I do not have time to get into that today." Pauli said looking at his pocket watch.

"And just when the conversation was getting good!" Jung joked as he gathered himself to leave. The men shook hands and Jung said, "I look forward to our next visit."

"Uf Widerluege, Doctor Jung." Pauli said, as he walked around the corner to a small train car that would carry him up the steep hill to the University.

FACTS

65.   It is believed that crows experience some form of conscious. They can think about the choices they make instead of relying solely on instincts. [Ref 58]

66.   Carnegie commissioned Napoleon Hill to write a book about the method he followed to accomplish his goals, becoming ultra successful. The process outlined in the book is suspiciously close to the law of attraction. See *Think and Grow Rich* by Napoleon Hill. [Ref 59]

67.   The time Pauli broke the joke was a real example of the Pauli Effect. At a reception, his colleagues tied a chandelier that would crash to the ground when they released a rope. When he walked in, however, the rig malfunctioned and the chandelier held firmly in place, thus becoming a real-life instance of the Pauli Effect. [Ref 60]

The Pauli Effect always seemed to work out in Pauli's favor and he always enjoyed when it happened. Another example is when he went to lunch with two people and as they stood up, they realized they had both sat in something but Pauli was clean. Naturally, the two blamed the Pauli Effect and Pauli was thrilled.

Chapter 22

Archetypes Everywhere

"We don't see things as they are, we see them as we are."

-Anais Nin

Dr. Jung sat at his desk reading notes from that morning's client. She had a dream that replayed a real event, so Jung was contemplating whether the dream was literal or if that event represented something to her and seeing it in a dream was symbolic. Jung's thoughts were alone in the calm silence of his office, accompanied only by the metronome of his pocket watch, persistently ticking from his front pocket.

Down the hall, Jung heard his secretary's chair legs scrape the floor as she stood up to open her window. He realized how stuffy his office had become from the warm spring sun toasting the terraced roofs of the town. He walked over and pushed the window open from the bottom, latching the hinge, locking it in place.

A cool breeze gently rustled his papers. The refreshing wind reminded him of Pauli's dream. He said it started in a classroom on a warm day where a breeze just like this blew across his desk.

A minute later, his tranquility was interrupted by the clinic's front door opening and quickly slamming shut. That was odd, Jung thought. He heard his receptionist greet his next client whose footsteps seemed to rush straight passed her. Jung looked at his schedule and realized it was Pauli.

He must be in an excitable mood, Jung thought as the receptionist chased Pauli down the corridor towards Jung's office.

Pauli walked in like a man who was in a hurry to get where he was going even though he had just arrived. Before the receptionist caught up to him, he entered Jung's office and shut the door behind him. The wind gave it an extra push and it closed harder than he intended which shocked the quiet office, slamming in her face.

"What's the meaning of this?" Dr. Jung asked, feeling a combination of surprise and concern.

Pauli briskly continued on his trajectory to the lounge chair where he plopped down. At that moment the window Jung had just opened slammed shut, breaking one of the panes of glass.

Pauli sat upright in his seat. Jung and Pauli glanced at the broken window then locked eyes. For a moment they stared at one another in an intense gaze until the corner of Jung's mouth cracked into a grin. The grin turned to a large smile then he started to laugh. Pauli looked from Jung to the window and realized the Pauli Effect struck again.

Pauli chuckled which grew into laughter until whatever stress he walked in with was laughed away. His shoulders relaxed and he said, "I have come to enjoy the Pauli Effect manifesting itself."

Jung laughed. "That was an entrance I won't soon forget! Now then, I was just thinking about that dream you told me about. Are you still having it?"

"Yes."

"I have been remiss in describing what I look for when I speak with my clients. Before we analyze your dream, I should explain the ideas of archetypes and heuristics."

"Archetypes like the persona and the shadow?" Pauli said.

"Yes, we have already discussed some, but the term archetype can have other meanings. Archetypes can be symbols within your internal psyche or symbols you recognize in the external world. Then there are heuristics which sound complicated but are just unconscious understandings of how something works."

Archetypes Everywhere

"How does someone understand something unconsciously?" Pauli asked. "For example, the 'light-from-above' heuristic. Light, generally speaking, comes from above us," Jung said gesturing upwards, glancing to the ceiling. "This is not a conscious understanding, it's something your mind automatically assumes. Light hits your eyes which send an electric signal to your brain and your brain creates the image of the things you see."

"Right, but where is the heuristic you are talking about?"

"When your brain is processing the data from the eyes, do you have to consciously figure out what all the shapes and colors are? Of course not, your brain unconsciously processes all the raw pictures and presents it to you."

"Okay."

"And the light-comes-from-above heuristic is a rule in that process. Your brain makes sense of the shadows it sees very quickly." Jung explained. "In contrast, when people tell ghost stories, they might shine light onto their faces from below. This creates an uneasy feeling that makes people feel off without being able to describe why."

"I understand," said Pauli, "if you had to figure out which direction the light was coming from every second of everyday, the brain would waste a lot of processing power."

"Especially considering the fact that light, almost always, comes from above so there is no point to figuring it out over and over again when it's the same answer almost every time. This unconscious understanding is a heuristic."

"Fascinating." Pauli thought aloud as he contemplated what other heuristics there might be.

"Heuristics are just an unconscious understanding that allows the brain to save time and energy as it's piecing together your internal understanding of the external world." Jung explained.

"Cultural biases must be heuristics then." Pauli suggested. "Something people assume to be true without even realizing they think it." [FACT 68]

"Right, if we hold a belief so strongly that it becomes a heuristic, we will begin to experience everything in life as if we were looking through the lens of that belief." Jung explained. "It's called confirmation bias. If you believe the world is a just place you will tend to see justice served. If you believe the world is an unjust place, then you will constantly notice every injustice to the point where you will begin to see injustices where they do not exist. That is all proven science." [FACT 69]

Pauli nodded, "Now what about archetypes? I already understand the ones pertaining to the shadow."

"The self, persona, shadow and the anima or animus are the ones I typically look for in my client's dreams" Jung continued. "When archetypes show up in dreams, they are symbols pertaining to one's own psyche but archetypes can be external to the individual as well. Patterns of human behavior, for example."

"Wait, what? I have several questions. First, what are the anima and animus?" Pauli was curious.

"The anima and the animus are archetypes that represent the aspects belonging to the opposite gender. So, a man has an anima representing his feminine qualities while a woman has an animus representing her masculine qualities."

"What is the purpose of that?"

"It's a tool our unconscious mind uses to highlight things about ourselves. It might be the nurturing aspect lacking in a man or a woman's overly logical way of thinking."

"Interesting." Pauli said. "You think those archetypes are somewhere in the dream I told you?"

"Yes. Archetypes can be shown to us in dreams as many things; symbols, animals, even people."

"These things show up as whole people?" Pauli was now intrigued. "People are complicated, how could one aspect of someone be an entire person?"

"It's called personification. Any emotion can be represented as a whole person." Jung explained.

"I am skeptical," Pauli said. "A whole character is too complicated of a product to just represent one thing. How are you so sure my dreams represent these archetypes you are thinking of?"

"Archetypes are not just in our dreams," Jung instructed. "All good stories have characters that follow a pattern of behavior, specific to one archetype at a time. The most popular story archetype is the hero's journey. All the best stories use the same patterns because we recognize the pattern without realizing it. The audience feels like they really understand the character." [FACT 70]

"I'm still not sure I understand what you mean by patterns of behavior." Pauli said, hoping for clarity.

"Have you ever heard of Romeo and Juliet?"

"Don't patronize me, sir." Pauli said calmly in his deep raspy voice. "Everyone knows Romeo and Juliet."

"Well Romeo is the personification of love. The audience relates to Romeo throughout the story because his actions show exactly how love would act if love was a man."

"Interesting." Pauli said softening his tone as he thought for a moment.

"Everyone in the audience has felt a deep infatuating love before. Since we are whole people capable of logic and many other emotions, we do not actually do what Romeo did. People see the actions of the character and recognize that specific part of themselves."

"So Romeo displays a pattern of behavior but ultimately, he is just a symbol." Pauli figured.

"Just like symbols you might see in your dreams." Jung added. "Take your dream for example. Your teacher might represent an emotion or a habit, like how dedicated you are to your work."

Pauli thought for a moment and realized, "the teacher in my dream might not represent my teacher even though he was a real person."

"The people we see in dreams might represent one single emotion like anger or freedom," Jung explained. "Then in the dream that person interacts with you in a certain way. For you, your teacher scolded you for not paying attention. If your teacher represents your devotion to work, then the dream might be showing you how that devotion is affecting other aspects of your life."

"My devotion to work is causing me to be chased by a dragonfly?" Pauli said sarcastically.

"The symbols don't have to be people; anything could be a symbol. Your dream might be telling you that your devotion to work is interfering with some other aspect of your life, represented by a dragonfly." Jung explained.

"Perhaps I was overthinking this whole thing." Pauli said as he began to understand the fluidity of dream symbols. "As the dreamer, I am the whole person, but each symbol could represent one emotion."

"What is more interesting is that over the years there have been at least three others who sat on that very sofa describing a dragonfly in a dream and one in real life that pestered them until they took notice." Jung added.

"There are many dragonflies in the world, of course others have dreamed of them at one point or another." Pauli said, totally glossing over the insane implication that a dragonfly randomly appeared in someone's real life and was somehow connected to their psychological health.

"Yes, but the interesting thing for each of you is that the dragonfly represents a similar hidden message. A message your unconscious mind is trying to tell your waking mind."

"How do you know?" Pauli challenged.

"For you, I believe it is telling you to acknowledge your anger."

"Let me get this straight. You are saying that some secret part of my own mind is trying to send a message to me through my dreams? And that message is that I am angry about something?" Pauli already understood the subconscious mind's role in this process but was abrasive because he was starting to face what his shadow was hiding.

"Yes." Jung said compassionately.

"I might find that hard to believe considering how easily I speak my mind at work."

"I don't think this has anything to do with your work Herr Pauli."

"Then what?"

"One of the reasons you originally came to me."

"My wife." Pauli sighed as his mind raced. "I do feel like the teacher represents my work in some way."

"I believe your shadow is the teacher telling you to focus on your work, hiding the truth." Jung explained. "Your dream's perspective is your persona."

"Wait, you are saying the me in the dream is not my whole self?"

"Think about it. Your conscious experience as you go around your waking life is not your self; it is your persona." Jung explained.

"A man's conscious identity is not his whole self, it is only his idealized self image." Pauli realized.

"That's right," Jung continued. "The dragonfly is trying to show you there is something you need to address, which might be your self trying to get your attention."

"I don't know about that." Pauli said folding his arms and leaning back in his chair, but Jung pushed on anyway.

"The dragonfly your self is trying to get your persona to acknowledge whatever your teacher is trying to distract you from but what that is, I have not yet figured out. Perhaps the second part of the dream has the answer. Water often represents your own psyche but what does a goose represent to you?"

"Hold on, are you saying *I* am the dragonfly attacking *myself*?" Pauli said, almost sounding appalled.

"Well of course, you are the dreamer of the dream, in this story there are no characters other than you. Every symbol is a different part of you; the chalkboard, the teacher, the dragonfly, even the classroom setting itself are all different aspects of you or how you feel about other people."

Pauli finally understood. The entire dream was Pauli's full self, the version of himself in the dream was only his persona. "The dreamer himself is broken into pieces, each disguised as a symbol within our dreams."

I don't want to push the counseling forward too quickly, Jung thought checking his pocket watch.

"Did I tell you about individuation?" Jung changed the subject just a bit.

"Come again?"

"Individuation is the process when someone such as yourself works to combine what their unconscious mind knows (Self) with the individual's public self-image (Persona)." [FACT 71]

"Individual- what?" Pauli asked.

"Individuation." Jung explained, "It's a process that can take time but each person must choose to consciously integrate their persona and self. You must acknowledge the bad parts of yourself which can be a very uncomfortable process. Once you do that the unconscious mind will no longer need to send messages in cryptic dreams."

"Then individuation is my new goal." Pauli said quietly.

The timer rang out.

"That is our time for the week. If you have the dream again ask yourself if any of the things I said could be true."

For a moment Pauli sat in contemplation looking at the floor. He finally took a deep breath and looked up, "allow me to use the restroom."

"Take as much time as you want my friend." Jung said.

FACTS

68.   The 'Light-from-Above Heuristic' is the actual name of that heuristic. As explained in the narrative, a heuristic is an understanding that our unconscious mind has for the world. When perceiving an object, our eyes see raw data such as shadows, shapes, and colors. The brain understands this rule as it is processing all that data to make sense of it.

Heuristics often involve aspects of our senses that we do not consciously process. They save calories and time by quickly gaining an understanding of the outside world. Heuristics are not limited to our senses. People have cultural lenses they view the world through, these too can be heuristics, though this is specifically referred to as cultural bias. Anytime someone has an unconscious understanding that something is a certain way; that is (or can become) a heuristic. [Ref 61]

69.   Confirmation bias is when people only accept information that supports their current understanding. They ignore data that might conflict with what they already believe, even if it is accurate data. [Ref 54]

We live in one of the safest times in all of human history and yet the average American will tell you things are far worse than they used to be. Why? Because we are constantly talking about the bad things which creates a belief in our minds. This understanding reciprocates and we start to notice all the bad things until we think the world is burning. Our society even has a doomsday clock; utterly ridiculous. [Ref 61]

We all see the world through our heuristics. If an individual is not aware of their cognitive biases or heuristics, then their shadows can grow. Confirmation bias can occur when the archetypal shadow becomes a heuristic, manipulating the individual's interpretation of the world. [Ref 63]

70.   An archetype is a standardized character or symbol that can appear in movies, books, dreams, and even real life. Jung focused on the four archetypes described in the narrative; the persona, the self, the shadow and the anima/animus but there are many more.

We, the audience, or the dreamers of the dream, unconsciously recognize patterns of behavior that the archetypal character or symbol represents which is how we relate to the characters unconsciously (emotionally). Like heuristics, archetypes provide unconscious understandings without a conscious explanation required. [Ref 63, 64]

71.   Individuation is a healing process in which someone learns how to reduce their shadow. Individuation is accomplished by combining the self and persona. In other words, the individual's goal is to openly live as the 'Self' in the objective world, without a persona at all. [Ref 63] This can only be accomplished if the person acknowledges they are capable of evil deeds. (*Capable* of evil, not *guilty* of evil)

We are not the ideal people we think we are. As a species, we are inherently selfish, and that's generally a good thing but can be taken too far. We are all willing to work for ourselves, so when we have a value-based economy where value creation is synonymous with personal wealth then the individual and society will flourish. In addition to being inherently selfish we are also social creatures who evolved to want to help others. People who help others tend to be happier than those who do not.

Individuation requires the individual to fully acknowledge their dark side. Whether we are aware or not, each of us is capable of great violence. If we weren't we would not survive. We must eat and the only way to eat is to kill something, even if just a plant, we must kill to survive. By pretending you are not capable of killing, creates an internal division between your real self and your ideal persona. There's nothing wrong with wanting to survive.

Chapter 23

The Entangled Webs
We Weave

"The total number of minds in the universe is one.
In fact, consciousness is a singularity phasing
within all beings."

-Erwin Schrödinger, PhD Physics
Pioneer of Quantum Mechanics

Pauli washed his face and returned to Jung's office refreshed.

"Remember those fundamental particles I told you about?" He asked from the same lounge chair he just spent the last hour in, though now that they were discussing physics, he felt much more comfortable.

"Electrons, protons and," Jung tried to recall, "neutrinos, was it?"

"Neutrons is what goes with protons and electrons but that is incorrect."

"Oh?" Jung puzzled.

"Electrons orbit the clump of protons and neutrons in the center of each atom, but protons and neutrons are not fundamental particles. Quarks are quantum level particles that form-"

"That form the protons and neutrons." Jung interrupted. "Yes, it's difficult to keep it all straight."

"Very good. Remember, there are other fundamental particles I purposefully left out of our previous discussions. All fundamental particles play key roles in our reality but they do not necessarily build the stuff of the universe, which was the focus of that lecture."

"That's right, there are force-carrying particles that are pure energy." Jung said confidently.

"I don't want to confuse you but I must clarify something. All the fundamental particles have mass and therefore are not pure energy. What I meant was that only quarks and electrons combine to form all the stuff that we interact with in human-reality."

"So even the force-carrying particles are actual stuff."

"All but two; the photon and the gluon. The photon is the fundamental particle that carries the electromagnetic force." Pauli said.

"Yes, I vaguely remember that discussion. The photon generates light and radio waves and whatnot." Jung stated.

"Not generates- it *carries* those things. The photon *is* light and radio waves and all those other things on the EM spectrum."

"Yes of course, sorry, I am just getting warmed up." Jung said as he reminded himself of the specific meaning of each word. Talking with Pauli was like crawling through a mine field of terminology and the only way to make it out was to use the correct verbiage.

"Let's discuss some quantum weirdness we see in these fundamental particles." Pauli began his lecture. "We've taken a laser and split off pairs of individual photons. When that happens, the pair become entangled."

"Tangled?" Jung said, intrigued.

"EN-tangled," Pauli corrected, "they are connected by some unseen means that we do not fully understand. You can physically separate the particles, even to opposite ends of the universe but this connection would still exist." Pauli explained.

"What is this connection?" Jung asked excitedly. "Do they communicate with each other?"

"No." Pauli said flatly. "Particles each have many properties and aspects that make them unique but I have not told you about any of that yet."

"Property meaning what?"

"Just look at your shoes for example. A property is a way to describe them. What is one property of your shoes."

"Well," Jung thought, "they are black."

"Perfect. Color is a property that makes your shoes unique."

"But other shoes can be black too, so it isn't exactly unique."

"That's correct but if you define every property of your shoes, they will be very unique indeed."

"I see," Jung said, "particles have their own properties. You said one is spin, what are some others, I can't imagine they have colors."

"Hm, intuitive." Pauli chuckled, "One of their properties is color but you are correct because these particles are too small to actually have color as we think of it."

"Particles are colorless but have a property called color." Jung was confused.

"Yes, it has to do with the strong-force, having three possible states that, when added together, must cancel out." [FACT 72]

"I'm afraid you've lost me." Jung admitted.

"You don't need to understand all that just yet." Pauli said. "Truth be told we just needed to come up with a way to differentiate one property that had three options, some of which cancel each other out."

"Honestly I'm not sure if I understand, but keep going," said Jung.

"The property we need to discuss now is spin. The spin can only be measured as spin up or spin down, there are no other ways a particle can spin."

"Spin up or spin down." As Jung repeated, a lightbulb in his mind flashed on. "I get it, there are more than two colors and the property had more than two possibilities and people inherently understand colors so when you found a property with more than two options you used colors." Jung understood.

"Yes, something like that. The property is actually more like an electrical charge but with three possible states, instead of just the positive and negative we are used to."

"That is fascinating." Jung said thinking of the possibilities of a third electrical charge.

"Now going back to our entangled particles; if one has a spin property of up, then its entangled partner *must* have a spin of down, since they are entangled, they must be opposite. Here is the rub. We do not know which particle has spin up and which has spin down until we measure one.

The moment you have measured one, you instantly know the measurement of the other, even if that particle was at the opposite side of the universe. That is the spooky action that Einstein described, you see?"

"Why is this so strange?" Jung challenged. "Clearly, when the particles were entangled, one was spin up and the other spin down. You just didn't detect which one was which until later. I don't see what is so spooky about that."

"Wrong." Pauli said, gently shaking his head. "They simultaneously exist in *both* states."

"How?"

"You are ready for the damned cat."

"Finally going to let the cat out of the bag, are you?" Jung exclaimed jokingly, seeing no possible connection to a cat.

"In the quantum age, Herr Jung, the cat is out of the bag but has found itself in a box." Pauli said mysteriously. "Schrödinger thought up an experiment where there is a locked box, inside of which is a cat and a vial of poison. An observer will open the box at a certain time to discover the cat's fate. Either the poison has broken open and the cat is dead, or the poison did not break open and the cat is alive. We do not know until we open the box and see." Pauli paused.

"This may be a little cruel but I still don't see anything spooky." Jung said defiantly.

"Give me a minute." Pauli said dramatically. "If you open the box and see the cat is dead you probably think, well at some point it died, then it sat dead inside the box until we opened it. That is not what's happening. Before we open the box, the cat is in a state where it is both alive and dead at the same time."

With Pauli's words, Jung was transported to a time almost 20 years ago when he was in his field office outside the little village on the edge of the Swiss Alps. He was speaking to Rupert about how he had entered such a calm state when he was in peril.

"It's like you were in a state of mind where you were both alive and dead at the same time." Jung could feel his heart rate increase. He blinked and brought his attention back to the present moment in Zurich with Pauli.

"I believe, now more than ever," Jung said, "that our two very different fields of study are, perhaps, not so different after all." His mind raced to understand what Pauli was saying.

"It's called superposition. You could think of it as potential."

"Potential." Jung repeated in awe with no idea what it meant.

"It's when the particle exists in both states simultaneously until it is measured." Pauli continued.

"It has the potential to be either." Jung noted.

"In the exact moment you opened the box the superposition broke down into one definitive state. Before you looked, the cat was, truly, both alive and dead." Pauli explained.

"How could that be true? If the cat was in both states at once, what would it think was happening, what would that experience be like?" Questions flooded Jung's mind as he tried to understand. Can people live in superposition? He wondered, thinking of Rupert.

"The cat is just an example of a quantum particle. The cat's possible properties are alive or dead; for the particle it's either spin up or spin down. While in superposition, it is both. We only know of this type of thing happening at the quantum scale so there is no need to hurt your head pondering what that would be like for us to experience firsthand."

"Okay." Jung sighed.

"Now, back to our entangled particles that are on opposite ends of the universe; both are spin up and spin down until we measure one. Once we measure one, we instantly know the spin of its partner, even though its partner could be billions of light years away."

"Wait," Jung said, "the particle we didn't measure, the one on the opposite side of the universe; what happens to it when we take the measurement of ours?"

"That particle instantly obtains the spin opposite its mate. They have a combined wave function which collapses when one is observed." Pauli explained.

"So, if someone was watching our particle's twin at the exact moment we measured ours, it might appear that the particle randomly changed right before their eyes? When in reality, it changed because of our actions on the opposite side of the universe?"

"Theoretically yes, but," Pauli tried to explain that technicality if someone was watching the particle on the other side of the universe that person would have already caused both particles to choose their unique state, but Jung couldn't contain his excitement.

"Then there *is* some unseen connection! What did you call it? Spooky action?" This is exactly what synchronicity requires to be true, Jung thought to himself excitedly, an unseen connection that exists outside of time and space.

"Quantum entanglement," Pauli specified, "is the preferred term. Spooky action at a distance was actually meant to criticize the theory."

"So, quantum physics has proven that we can affect things in unknown locations from great distances." Jung surmised still wrapping his mind around the implications of entanglement." [FACT 73]

FACTS

72.   The Pauli Exclusion Principle, which is the work he won the Nobel Prize for, states that each particle in a system must have unique properties. Color is one of these properties. They chose color to describe this property, not because the particles literally have colors, but because it is a description we intuitively understand. If we add colors together, they become different colors just as particles add and cancel out each other's color properties. The color property is more like an electrical charge than colors.

73.   The description of entanglement being a physical connection of any kind is not how modern science conceptualizes it. Mainstream scientists have no idea how entangled particles interact. The assertion that there is some, currently undefinable, but literal connection is an intuitive assumption made by the author.

Scientists insist that no information has traveled faster than the speed of light even though entangled pairs resolve instantaneously regardless of how far apart they are. In the narrative, entangled particles were described as having a combined wavefunction; this is another assertion made by the author which would explain how particles can display this instantaneous behavior.

A pan-psychism reality would intuitively explain the entanglement conundrum that science currently has no meaningful explanation for. If the combined wavefunction that entangles particles is some form of consciousness then that same consciousness could be the universal 'observer' that assists in the process of waveform collapse. (More on this later)

Chapter 24

Superposition is Specifically Unspecified

"Quantum entanglement is not all," Pauli lectured, "I must explain how a particle behaves in superposition and the inherent misunderstanding of the observer's influence on the measurement."

"That sounds complicated." Jung said.

"It will make perfect sense to you by the end of the lecture." Pauli reassured. "Later we will discuss if a tree makes a sound when it falls in the forest and no one is around to hear it, but first we will look at different wavefunctions."

"Are those more like a ripple on a pond or a wave at the beach?"

Pauli didn't understand the question. Ripples on a pond are just waves seen from above. If you change nothing but your perspective by standing on the same plane as ripples, then they look like waves at the beach. "I'm talking about wavefunctions, not waves, but I will answer that in a minute when we discuss the double slit experiment." Pauli said authoritatively.

"Alright." Jung accepted.

"Wavefunctions are mathematical expressions that can represent all kinds of things." Pauli explained. "We already established that superposition means the particle is in multiple states at once, but also, it is nonlocal."

"It's from out of town I guess." Jung joked. Pauli glared at him with a dry, unenthused expression as if he was silently scolding a child.

"Local, meaning it is not in superposition and it has a definite location. Nonlocal meaning it is in superposition and we don't know its location." Pauli explained.

"We can only guess where it is, then." Jung figured.

"No, we can calculate," Pauli corrected. "The question is, what are we calculating? We are not looking for its exact position because it's in superposition, it literally does not have an exact position."

"Potential!" Jung said remembering the previous explanation.

"Yes," Pauli said, almost surprised Jung seemed to be keeping up with the lecture, "that's correct. For this exercise we will calculate the particle's *potential* location."

"How do you calculate something like that?" Jung asked.

"Well, you and I won't be doing any calculations, instead I will draw some graphs that represent calculations, this way we can still understand the physics and skip all the math."

Pauli found paper and a pen then drew a straight, horizontal line and labeled it, 'Line of Existence.' Next, he drew what started to look like a bell curve squished together resembling a spike.

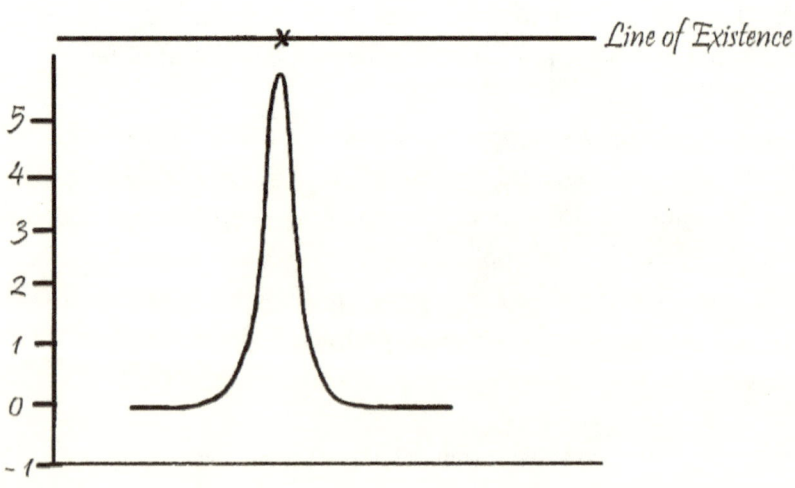

"This is a waveform." Pauli said. "It describes the potential for where the particle is. Please indicate where you think the particle is based on this graph."

Jung was unsure what information he was missing. He looked at the spike and figured it must be in there. "Anywhere under that arc, I suppose."

"Ah, I did not give you proper instructions." Pauli realized. "The particle is somewhere on the 'line of existence.' The graph below the line corresponds to the particle's energy at each point along the line."

"Does that actually happen in nature, where particles are confined in some way?" Jung asked to clarify.

"It does." Pauli said with a nod. "When electrons are part of an atom they can only exist on that atom's orbital. Free floating electrons, on the other hand, would not be confined."

"Orbital rings hold electrons around the nucleus of atoms." Jung remembered.

"Think of probability, when an electron is in an orbital there is a 100% chance it is on that orbital plane, however, it could exist absolutely anywhere on that orbital. It is just energy and if a second electron is added then both are just energy waves spread out over the entire orbital."

"Alright." Jung said as he prepared himself for the mental exercise. "But before I do this, tell me, what does this spiked curve represent?"

"This waveform represents a localized particle's energy, spread out over the line," Pauli said.

"It doesn't seem very spread out." Jung noted.

"This is a localized particle."

"Not in superposition then."

"Correct," Pauli said getting impatient, "now please indicate the spot on the line where it is."

"This seems a bit ridiculous, clearly it is here." Jung said as he drew an X on the line just above the top of the steep arc.

"You are correct. That was an easy question designed to help you understand the next few." Pauli said as he drew a new graph and 'line of existence.'

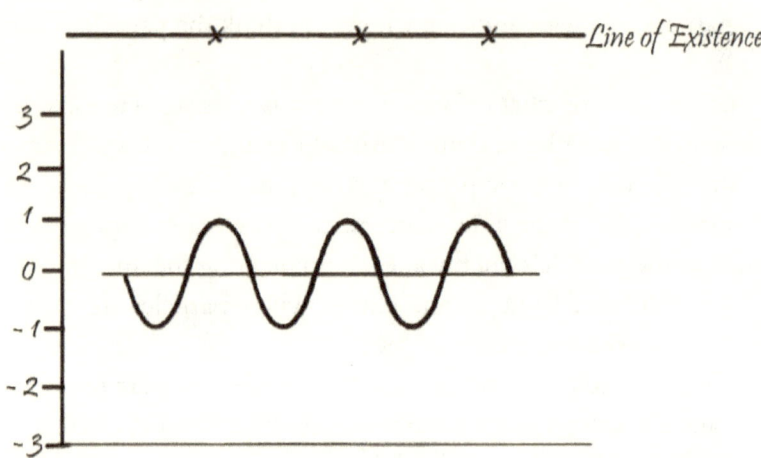

"This is a waveform that represents one photon by itself." Pauli explained. "Please indicate where on the line of existence the photon is."

Jung was thoroughly confused. There was no point higher than any others. "I suppose in any of these locations." He guessed, drawing Xs on spots above each wave crest.

"You're on the right track but before I give you the answer, let me draw another graph so you can see the answer to both." Pauli said happily.

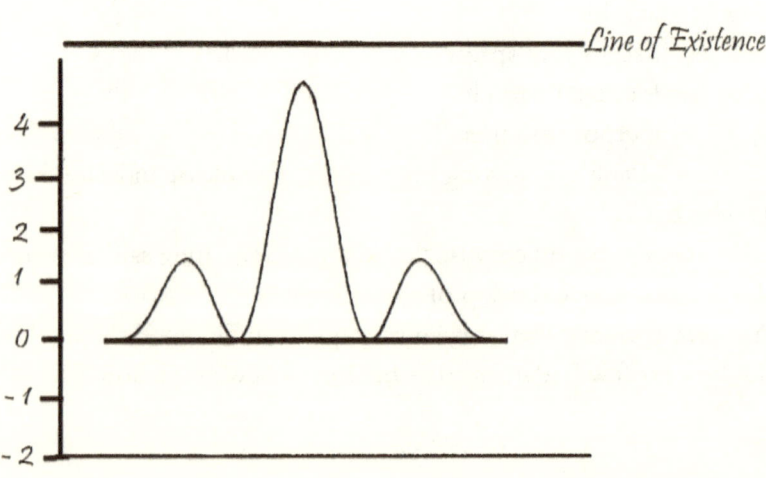

Superposition is Specifically Unspecified

Jung wasn't sure what Pauli was enjoying more; this game or Jung's frustration with it.

"This waveform represents a photon that has been influenced by another. It's called an interference wave." Pauli explained.

"How would the two photons have interacted?" Jung asked.

"They can in many ways. For this one let's say they combined."

"So, this is two photons?"

"Not to confuse you but this is actually still just one photon." Pauli explained.

"Okay." Jung puzzled.

"I'll explain everything in a second." Pauli said. "Just tell me where the particle is on the line." He instructed.

Jung felt like he was walking headfirst into a trick question but he didn't know what he didn't know. "I suppose it would be here, where the wave is highest." He guessed.

"That is where the particle is most likely to localize at this exact moment in time," Pauli said, gesturing to the graph, "but it is not where the particle *currently* is."

Jung thought for a moment. "There is a lot of implication in that statement. For one, you are saying this odd graph you drew changes over time. The particle only looked like this for a moment then what happens?"

"Very good Dr. Jung. The particle is a wave of energy so it would still be a wavefunction, but it would be flowing like when you pluck a guitar string, the waves vibrate the entire string at different amplitudes but you only hear one sound."

"If the highest point is where it is most likely to localize but the highest point is shifting in time then the most likely point where it would localize also changes." Jung figured. But where is the particle *before* it localizes, he thought.

Jung compared the two wavefunctions Pauli had drawn with this third one. The first curve intersected with the zero line almost immediately and never crossed back to be positive. The second two graphs were wider. They crossed over the zero line a few times only combining with zero at the far edges.

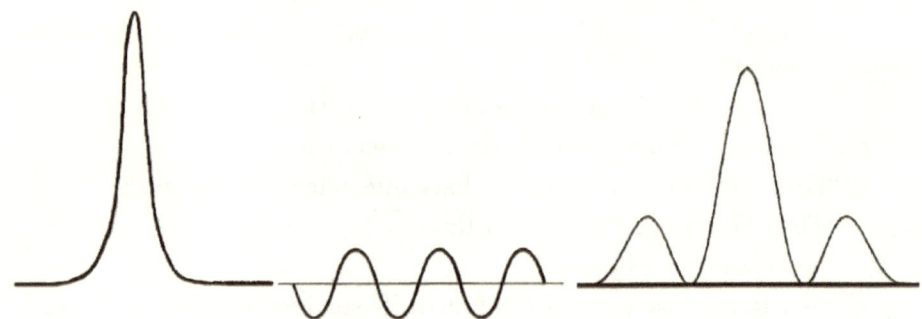

"I've got it!" Jung exclaimed. "The particle simultaneously exists in all places on the line where the wavefunction is positive."

"Absolutely correct, but how is it in more places than one?" Pauli encouraged.

"It is in superposition. Nonlocal and we are not calculating its position, we are calculating its potential to be in any one place at this moment!" Jung understood.

"Brilliant!" Pauli exclaimed.

"What about the places where the wave goes below the zero line?" Jung wondered.

"For this snapshot in time there wouldn't be any energy in those locations, however these lines are two-dimensional representations of a three-dimensional thing."

"Why reduce it then?" Jung asked.

"Well, I haven't got a three-dimensional piece of paper now do I?"

"Right," said Jung.

"We reduced it to two dimensions so we can easily learn about it but when we do that some data is lost or misconstrued. The point of this was to show you that particles' energy can be expressed mathematically as a wavefunction." Pauli finished.

"And a particle exists across the entire space where it *could* localize." Jung added.

"Now I can explain how that one photon was interfering with itself to create the last wavefunction I drew." Pauli said.

"Yes, that was very odd indeed."

"It all started with the double-slit experiment." Pauli began.

Dear Reader,

If you are familiar with the double-slit experiment and how waves from a single photon create interference patterns, then feel free to skip ahead to page 273.

"The very first double slit experiment was done by Thomas Young back in 1801." Pauli began.

"Quantum physics started that long ago!" Jung exclaimed.

"No." Pauli said sharply as if he purposefully wanted to sound dry and snooty. "Young proved light acts like a wave. It was much later, we realized it acts both as a wave and a particle, meaning it had a quantum explanation." Pauli explained. "The experiment goes like this; there is a light source that shoots single photons or electrons at a detection screen on the other side of the room. I will attempt to draw this."

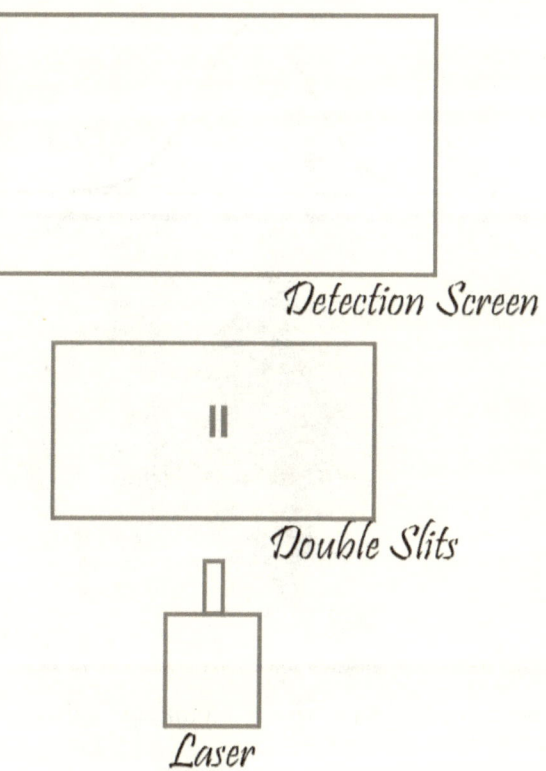

Detection Screen

Double Slits

Laser

"In the middle we put a shield that blocks all the photons except for the ones that go through the two tiny slits in the middle. This way we know that if a photon reaches the detection screen it must have gone through one of these slits."

"Okay, simple enough." Jung said.

"We shoot one photon at a time and record where on the detection screen it hits. If we repeat this many times, what kind of pattern do you think will develop on the screen?" Pauli challenged.

"Well, I would think most would hit closest to the center then dissipate moving outward." Jung reasoned. "But I assume that is not what happens."

"Let me draw what you described with an associated distribution curve." Pauli said, adding data.

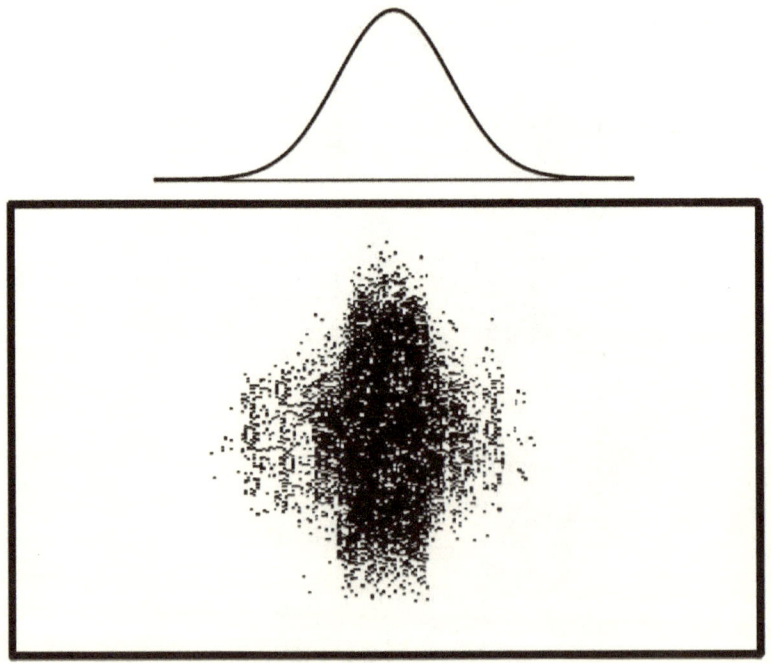

"Is this more or less what you are thinking?" Pauli asked.

"Yes."

"This graph above the dotted screen is a measurement of the screen. It is called a normal distribution and it's what you might expect. Most photons hit near the center and as you get further away from the center there are less hits. But remember there are two slits the photons could pass through."

"Right. Then I suppose it would be like two of those curves next to each other, like this." Jung presumed as he scribbled a crude sketch.

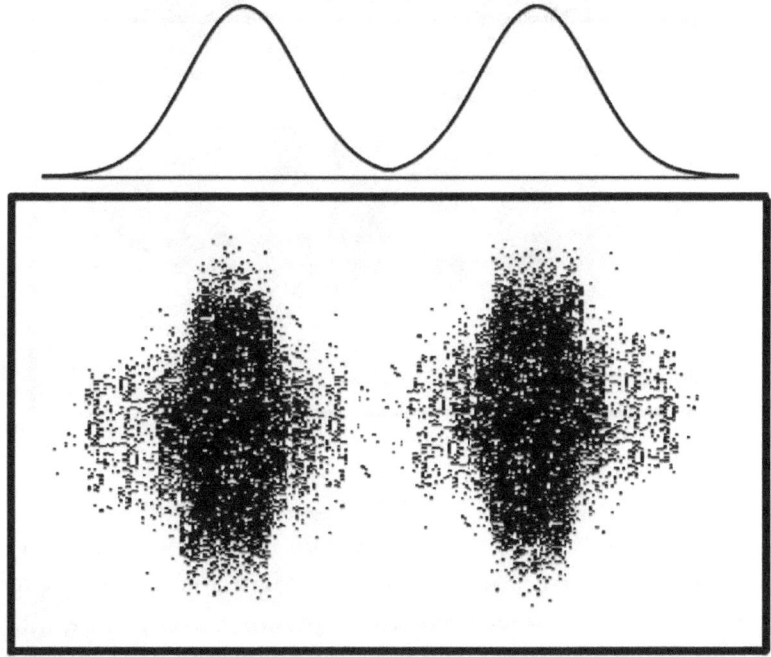

"This is what we would expect if the particle was a particle and not a wave. Another reasonable guess, but no." Pauli said. "What we see when we actually run the experiment is an interference wave."

"Like from before?" Jung asked, as he recognized the wave as Pauli drew it.

267

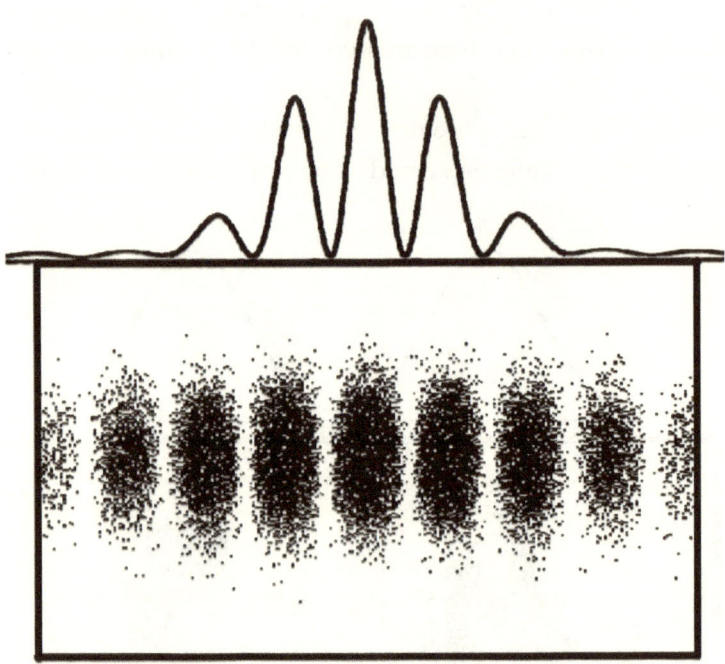

Jung studied the pattern, comparing it to the graph. He lined up the tops of the waves with the columns of dots on the screen. "I get it now!" He said, looking back at the normal bell curve and its associated dot pattern. "Where the waves are positive is where the photon could localize and where its negative it has no chance."

"Now you've got it." Pauli exclaimed.

"But what causes this pattern?"

"When energy in waveform combines with other energy in waveform it interferes."

"Right, you've already said that." Jung said impatiently.

"Let me show you."

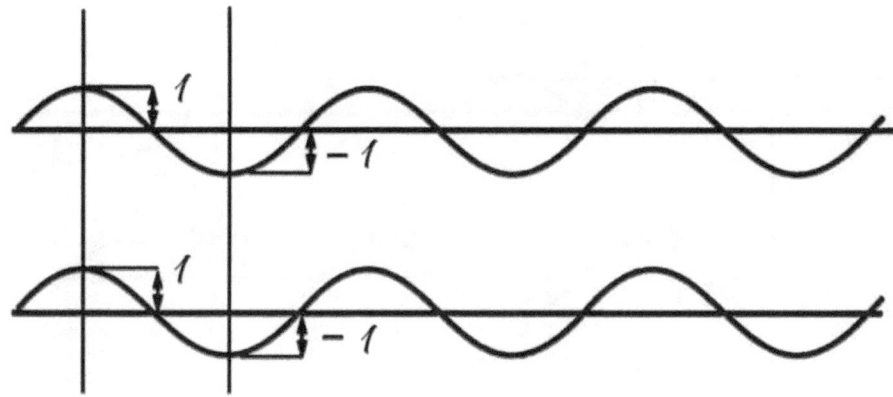

"Here are two general waves, what happens when they combine?" Pauli asked but Jung stayed silent.

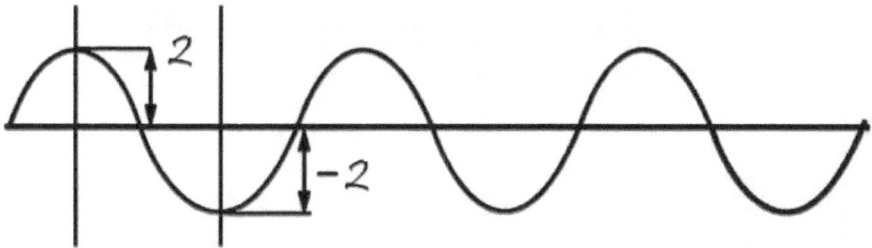

"Wherever positive energy meets positive energy, it adds together and when negative meets negative it also adds together but because it is negative the waves get lower." Pauli explained.

"Right." Jung said, remembering the previous conversation on the strength of the waves.

"Now what happens when one looks like this," Pauli began to draw, "and the other looks like this?"

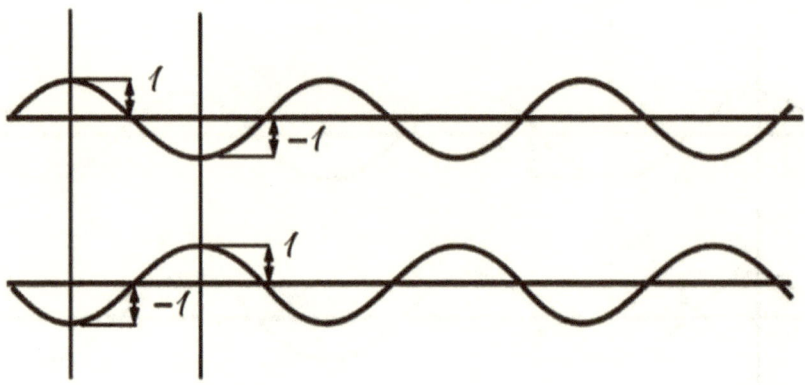

"I suppose the photons would repel each other." Jung guessed.

"Good guess, but in this case, they would cancel out."

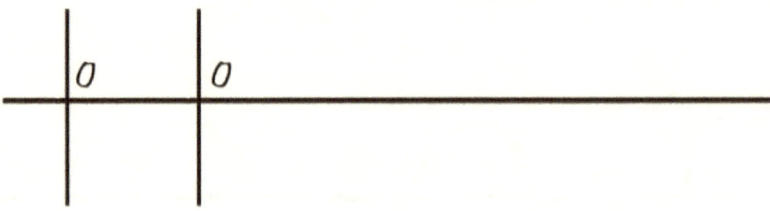

"Two photons could cancel each other out?" Jung was shocked at this realization.

"No, photons do not cancel each other out. I am showing you energy in waves and how the energy adds or negates to the combined wave." Pauli explained.

"I'm not sure I understand. I thought we were discussing photons." Jung said in frustration.

"Actually, we are just discussing one photon." Pauli said with a smirk.

"Now I'm thoroughly confused. How could one photon interact with itself... as waves, no less."

"Let me show you what is really happening with the double slit experiment." Pauli said.

"This is the one photon after it passed through the slits. It is still a wave because it has not localized yet." Pauli explained.

"Wait, which slit did it go through?" Jung asked.

"Both." Pauli said with a smile.

Before speaking again, Jung thought for a few moments as Pauli continued to draw waves that began to overlap.

"One single photon split into two, passes through small holes, then recombines?" Jung asked.

"The particle doesn't split. The slits must be extremely small and extremely close together for the experiment to work because the spacing must allow for one wavelength of light to hit both slits simultaneously. Do you know what each of these arced lines represents?" Pauli asked abruptly.

"A wave." Jung said.

"Almost. It is the crest of the wave." Pauli corrected.

"Ah, just the top point of each wave." Jung understood.

"Perspective is everything. On the left we are looking at waves from above. Right about here," Pauli said drawing a line just in front of the slits. "The waves would look like this if we were looking from the side."

"Instead of from above," Jung understood.

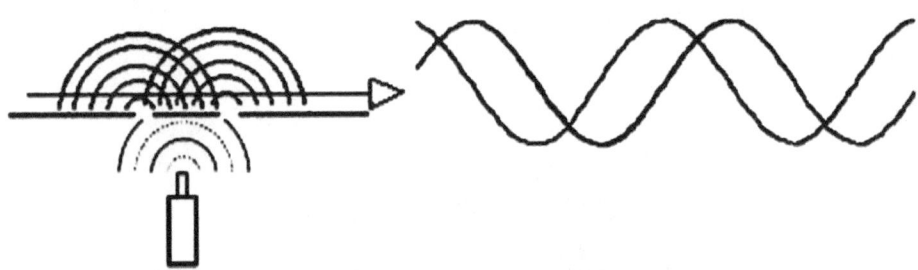

"Let's add it all together!" Pauli said excitedly. "Immediately after coming out of the two slits the waves are offset, like this. The two waves overlap, meaning what?" Pauli challenged.

Jung studied the drawing for a moment. "They get stronger where they both have positive energy and they get weaker where negative."

"Correct, just like this." Pauli said as he combined the wavefunctions.

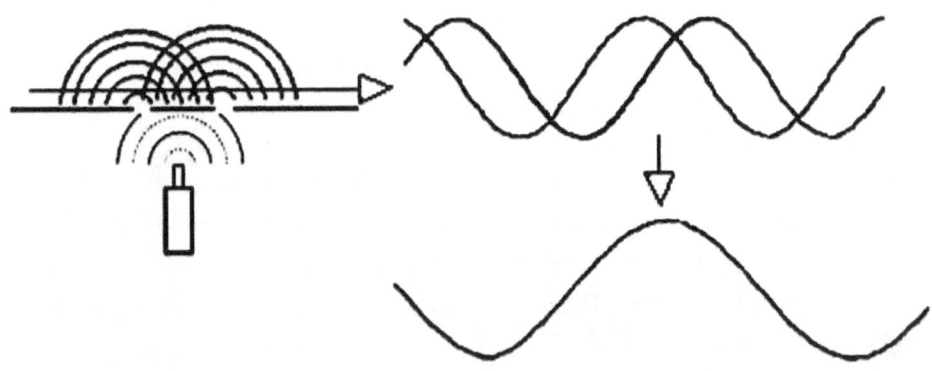

Pauli continued, "as the waves flow away from the slits the interference gets stronger." Pauli drew another line further out. "This is what happens when the interference pattern compounds as the waves ripple out."

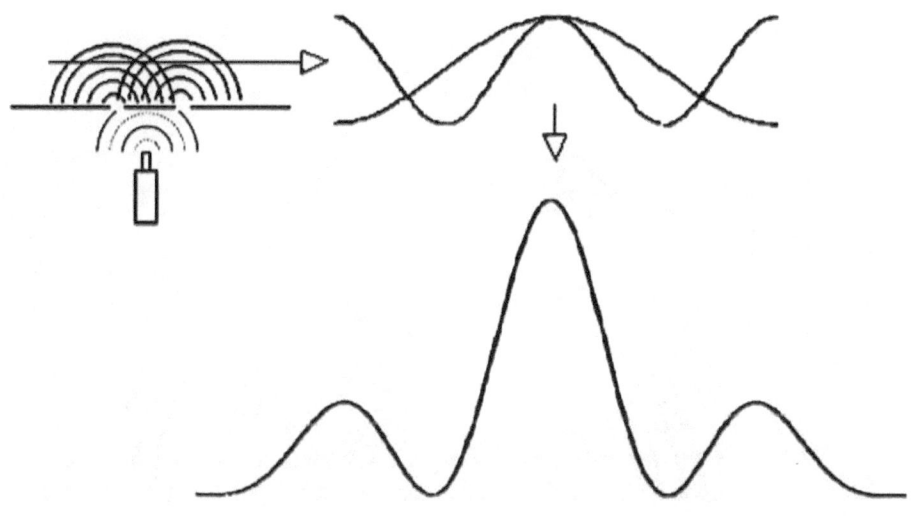

"Here we have one new wave and one that has already combined with another." Pauli said.

"I recognize that wave," Jung said, "that is one you already drew."

Dear Reader,
If you skipped this section, start again here.

"Very good." Pauli said as he completed his drawing. "This is the overhead view of how the light is traveling, and interfering, between the slits and the detection screen."

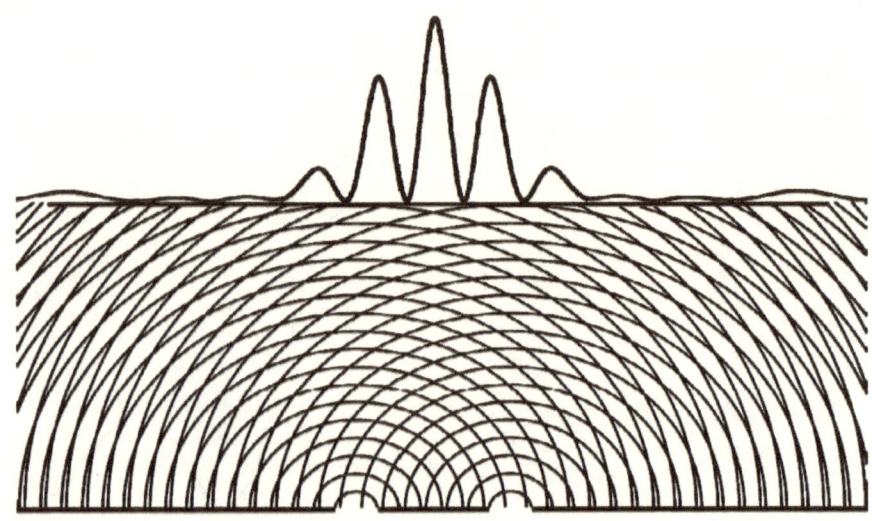

"And this is what the detection screen actually looks like because of the interfering waves." Pauli culminated after finding a previous illustration.

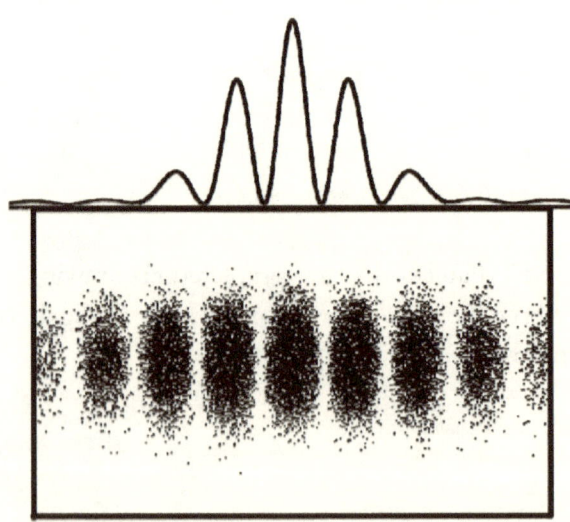

"Which proves that light acts as a wave," Jung said before adding, "sometimes!"

"Only before each particle localizes. After waveform collapse it acts like a traditional particle." Pauli explained.

"What does this all mean?" Jung asked.

"If a tree falls in the woods and there's no one around to hear it, does it make a sound?" Pauli asked.

"I am familiar with this one. If you believe in realism then everything is real, meaning the tree makes a sound. If you believe in nonrealism then all of reality may require a conscious observer to experience it, therefore if something happens where there is no conscious observer then it didn't actually happen." Jung explained.

"That is the philosopher's understanding." Pauli began, "The physicist would ask, does the thing have a definite value even if it has not been measured?"

"What does that mean?" Jung asked.

"If the tree adheres to realism, then not only would it make a sound but we could calculate a value, like 100 Hertz, using indirect measurements in the calculations. If the tree does not adhere to realism, then we can only know the value of its sound if we measured it directly." Pauli said.

"But nonrealism seems nonsensical." Jung began. "We know reality is real and the tree would have made a sound, regardless of an observer."

"Not so fast," Pauli said gleefully. "What we have realized is the answer to that age old question might be the tree did not make a sound after all."

"How could it not?"

"This is another analogy like the cat in the box. Ultimately what you must understand is that I am talking about a quantum particle in superposition, not a tree in the forest." Pauli explained. "The tree is an extrapolation, meaning if realism is not true for a particle, then nothing is real unless it is measured."

"I'm not sure I understand." Jung was hoping for clarification.

"Einstein does not like this one either, but unfortunately for my dear friend, he has been proven wrong. His argument was, is the moon not there unless you are looking at it?"

"Okay." Jung said feeling uneasy about what the answer might be.

"With the knowledge of Bell's Inequality Theorem, we now know that when a particle is in superposition it has all possible properties." [FACT 74]

"Like spin up and spin down at the same time." Jung interjected.

"And its exact position and velocity cannot be known…" Pauli continued.

"Right, we discussed all that already." Jung was becoming frustrated.

"How much energy it has is also undefined." Pauli continued.

"Meaning we can't measure it until after it localizes?"

"It is more than that. It means that while in superposition each particle has an undefined amount of energy." Pauli explained.

"The particle in superposition does not have an exact amount of energy? Then it really is all potential when it's in superposition." Jung said, not sure if he fully understood the implications. "It's almost as if particles in waveform don't exist but after they localize, they do exist."

"Particles in waveform do exist but only as undefined packets of energy." Pauli explained.

Jung thought through the implications of this and said, "If you can manipulate what a particle will become, or how it will localize, via entanglement then you could influence things over great distances and there would be no trace of your influence!"

"Theoretically that is correct Dr. Jung but as of now we have no control over the properties a particle obtains upon localization."

"The moment a particle localizes seems very important now. I know you said the particle exists while it's in superposition but it feels like what you are describing is that localizing is the exact precipice when a particle is definitively in our reality." Jung pressed, "how do particles localize anyways?" [FACT 75]

FACTS

74.   Bell's Inequality Theorem has been called the strangest theorem in science. It requires that either things can move faster than light or the universe is fundamentally, or at least has the capacity for, nonrealism. The prevailing thought is that nothing can go faster than the speed of light.

Therefore, our objective reality must, in some way, allow for nonrealism. This has led to many interpretations such as if a tree falls in the woods, it only makes a sound if there is a conscious observer. Spoiler alert, falling trees make a sound even if no one is there to hear it. This theorem will be interpreted later in the book. [Ref 66]

75.   The scientific explanations of superposition and entanglement described in this chapter are scientifically accurate to modern theories. [Ref 67]

Chapter 25

Reality is Only as Real as You

"Those who are not shocked when they first come across quantum theory cannot possibly have understood it."

-Niels Bohr, PhD Physics
Pioneer of Quantum Mechanics

"Waveform collapse," Pauli lectured, "otherwise known as when a particle localizes."

"Waveform collapse," Jung stated, trying to understand the context of all those drawings. "All the graphs you drew, those were particles when they were acting like waves but now you are going to talk about when they act like traditional particles."

"Actually, I'm talking about the moment they become traditional particles," Pauli said. "*When* is waveform collapse? For our example of the cat in the box, it was the moment you opened the box."

"Before we open the box the cat was both alive and dead and based on what you just explained we could also say it was both fat and skinny. Truely a wave of potential. After we open the box, we force one outcome to be true, the wave is gone and we have a normal particle with one specific amount of energy." Jung understood.

"And remember, the particle as a wave is spread out, physically existing everywhere it could exist; in all possible points in spacetime. Then the waveform of potential collapses and the packet of energy becomes a real, traditional particle that can fully interact with other things in reality."

"Hang on, there is a huge implication to that statement!" Jung realized. "Last week you told me that mathematically the future could influence the present and that space and time can warp. If a particle literally exists simultaneously at every possible point in space-*time* then part of it might exist in the future!"

"You are theoretically correct, but as of today we have no observations of this in nature." Pauli stated.

"How could you?" Jung lost himself in a thought. "What is consciousness and what created reality?" He mumbled.

"Pardon?"

"Yet again, I find myself at one of my two core questions. If consciousness exists outside the physical body, in any form, then it could exist in the future. And with what you've just described I can't help but wonder if my future self has any influence over my present self."

"A very philosophical question indeed. I'm afraid I cannot help with that one." Pauli said.

"Nonsense!" Jung exclaimed, "you are helping grandly."

"What was your other core question?" Pauli asked, setting a trap like a wolf driving its prey to an ambush.

"What created the universe?"

"That is an immature question." Pauli said bluntly. "A better question would be what is creating the universe."

"What do you mean? That's the same question!" Jung became frustrated.

"And yet it is fundamentally different."

"But I don't understand. What is *creating* the universe? The only difference in our questions is based on timing but the answer should be the same to both questions so why make that distinction?"

"You are so wrong your ignorance can't recognize itself." Pauli scolded.

"Then perhaps," Jung's eyes got big as he spoke as sarcastically as possible, "you could enlighten me with your grand and omnipotent wisdom." "Your question is asking how the universe was created, the answer being some sort of event, possibly the Big Bang. Then the universe existed through linear time. My question," Pauli spoke dramatically, "gets straight to the heart of the hidden knowledge you seek." Jung disregarded Pauli's tone and listened to his message. "Our reality is being created in real-time during every 'present' moment."

"I thought you said the past and future can equally influence the present but now, what are you saying? There is no past or future?"

Pauli's devilish smile crept up his face as he let Jung stew in the strange concept. Jung thought for a moment until Pauli finally spoke. "Have you ever experienced being in the past or the future?"

"Of course! I was in the past last week, last year and just a moment ago!"

"Ah, ah, ah." Pauli wagged his finger, "when you were experiencing your past, it was the present moment. Likewise, you can plan and imagine your future, but you cannot experience it. I think, therefore I am. Isn't that the idea you philosopher-types love?"

"I think, therefore I am?" Jung thought aloud. "Fundamentally I don't know that you exist but I know I exist because I can experience my own consciousness. You are telling me that the past and future don't exist because we cannot directly experience them?"

"They are concepts, but they can never be. There is only an eternal present moment."

Jung remembered a previous lecture when Pauli quoted Einstein in describing time as a persistently stubborn illusion. How could physics and math suggest that the future can influence the present and at the same time say there is no future or past?

Jung retraced the steps of the conversation, wondering how they arrived at this strange concept. He had asked his fundamental question; what created the universe?

"I think I understand." Jung said softly. "When philosophers and religious clerics discuss the creation of the universe by some higher power, they only discuss the beginning. Christians believe the world was created in seven days."

"Yes, I am familiar."

"But an equally perplexing creation mystery is of the present moment that is eternally manifesting itself as we experience it," Jung understood.

"It is the creation mystery of the quantum age." Pauli declared. "As you will find in our future discussions, the human reality we see is stable because of the energy constant flowing through the quantum fields." [FACT 76]

"The Tao." Jung mumbled questioningly. Pauli was jumping all over the place with these deep insights, Jung tried to stay on subject. "Back to waveform collapse. What causes it? And if the whole particle is spread-out what happens to the energy that is not near the point it ended up localizing to, it just disappears? How would it instantly become a single point?"

"Ha! Very good, Doctor Jung! Now you are really thinking! Particles exist in superposition until something forces them to localize. An honest physicist will acknowledge that exactly how and why particles break superposition and localize is still a great mystery."

"What do you mean something forces them to localize?"

"Ah yes, we are ready to discuss the inherent flaw I mentioned earlier. A particle can always be described as a packet of energy, either local, like a traditional particle that follows classical physics, or nonlocal. When the particle is nonlocal you now understand it is real energy in the form of a wave of potential."

"Yes," Jung said as patiently as he could, "a nonlocal particle is a packet of pure energy, mathematically described as a wave."

"Let me point out that even as pure nonlocal energy, the particle is physically real," Pauli said,

"Hm," Jung grunted as he pondered that concept.

"Here is the inherent problem; if scientists want to understand particles, we have to observe them."

"Okay." Jung said.

"But it is not possible to observe a particle while it's in superposition. The act of observing causes waveform collapse and might change aspects of the packet of energy before we can see what's really going on." Pauli explained.

"Are you saying the observer is actually influencing the behavior of the particles?"

"I know what you are thinking." Pauli said. "What you are thinking right now is the flaw I'm referring to. The particles do change because we are looking at them but that does not mean our consciousness is affecting them in any way. Particles are very small; do you agree?"

"Of course," Jung shrugged.

"Then we cannot simply look at them with our eyes as they glide past us, so how do we observe them?"

"I suppose with some fancy machines or computers."

"We use very sensitive detectors that record the energy when a particle hits them. The act of physically hitting the detector screen causes the waveform of potential to collapse out of superposition. The packet of energy is then 'observed' as a single traditional particle and it transfers all its energy to the detector, obliterating itself."

"Oh," Jung said.

"Here is an interesting way of looking at it; when a photon from the sun is absorbed by your skin, your skin is measuring that particle." Pauli explained.

"You are saying that particles usually travel in superposition but when they hit something, like a ray of sun hitting my skin; that localizes it?" Jung asked to clarify.

"Let me be very specific here, it is not enough to say the particle hit you. For the particle to be 'measured' it must transfer its energy to your skin, at which point, the particle no longer exists. Particles can bounce off things without causing waveform collapse as long as they don't transfer all their energy."

"When the particle transfers its energy, it no longer exists but if it doesn't transfer its energy then it isn't detected at all?" Jung repeated. "Why can't we detect it without stealing its energy?"

"Energy cannot be created or destroyed. The detection screen can only 'see' energy that has been transferred to it. Since the photon is not creating energy, it loses its energy by transferring it. If a particle bounces off

our detector it does not transfer enough energy, and we cannot detect it."
[FACT 77]

"This is what you meant when you said we change the particle by observing it." Jung realized. "Before we measure the particles, they exist in some unmeasurable state where they may be doing things we do not expect." Jung realized this uncharted territory of quantum science.

"In the case of the cat in the box, nothing is defining its property, whether it is alive or dead, until you look at it." Pauli said.

"Ah, I see how it works," Jung realized. "It seems like my consciousness is directly causing waveform collapse because I looked at the cat. A more accurate description would be the only way to see the cat is to obliterate the box, cat and all, then read the results on a computer."

"Very good Doctor Jung, that is correct," Pauli encouraged. "Remember, this is just a thought experiment for how we detect a particle localizing into one of its possible states, it is not how particles interact with each other."

"Got it," Jung said as he thought for a moment about how a particle in waveform is a definitive type, such as a photon, but all its possible properties like spin, color, frequency, and even its exact location and size are literally all possibilities at once. "When a particle localizes, it becomes one spin, one color, one exact location, but why those specific ones? Can those choices be influenced or is it destined?" Jung trailed off into a thought.

"Excellent question!" Pauli exclaimed. "We still have new information to go over before you will really understand just how profound the answers are, but what I will tell you is this. Different physicists have come up with their own interpretations of what influences particle properties upon localizing."

"So, no one really knows."

"The three most popular theories are the Copenhagen, the de Broglie-Bohm, and the Many Worlds Interpretations." Pauli explained.

"Yes," Jung said, "you told me about these before."

"The question these theories are trying to answer is how or why did each particle choose its specific properties upon localizing? Did the particle randomly become spin up or did something coax it into becoming spin up."

"Okay," said Jung realizing this explanation was going deeper than the last time Pauli talked about these theories.

"The Copenhagen interpretation requires that when a particle's waveform collapses properties are chosen at random like flipping a coin." Pauli explained. "Being random removes the possibility of a single destiny that we are all forced to live out."

"Is it truly random though, or could there be some underlying principle of consciousness that influences its outcome?" Jung asked presumptively.

"Don't get ahead of yourself Doctor Jung. The next interpretation is the de Broglie-Bohm Interpretation that postulates there are guide waves which, sort of, pull particles to their destined state. Meaning each particle is forced to become spin up or down based on the influence of these guide waves."

"Could the guide waves be influenced by something? How far into the future do the guide waves control?" Jung's mind was running wild.

"We are talking about the moment a particle localizes, literally just the present moment." Pauli explained.

"You keep saying spin up or down but what you are describing applies to all the particles' properties, correct?" Jung clarified.

"Yes. The last theory is the Many Worlds Interpretation which changes our whole perspective on what is happening when a particle is in superposition. 'Many Worlds' suggests that anytime a particle enters superposition, all of reality splits and all possible outcomes are literally true, just not in the same reality. There would be a whole reality where the particle is spin up and a totally separate reality where that particle is spin down; we just don't know which reality we are living in until it localizes." [FACT 78]

"That is absolutely fascinating, but which theory is correct?"

"Personally, I do not like the Many Worlds Interpretation. Just think of how much energy would be required to manifest a near infinite number of realities like that. Our universe seems to always use as little energy as possible."

"Like with our own physiology!" Jung added. "We evolved to be more efficient, so we use energy, thereby burning less calories."

"Right, and think of electricity, it always follows the path of-"

"Least resistance!" Jung interjected.

"Personally, I think the Copenhagen interpretation is most accurate." Pauli said, "the math seems to work out to indicate randomness."

"What about the particle's exact location?" Jung wondered. "When the waves of potential interfered and the ripples grew larger, we could predict where it would localize. That seems less than random."

"When particles are in superposition, we can often predict their location or velocity based on where the majority of their energy is."

"The energy is collected around one point." Jung offered.

"But sometimes when a particle's waveform collapses its exact locality seems to have been produced by random chance." Pauli explained.

"I am very intrigued by the mystery of why particles obtain the properties they do." Jung said. "You explained consciousness has not been seen influencing how particles express themselves but could it?" Jung paused to collect his thoughts before realizing he was back to one of his two core questions, so he added, "where do physical particles come from?"

"You may have heard of $E=mc^2$." Pauli said taking a breath.

"Yes! That is the formula everyone was making such a big fuss about a few years ago."

"Mass and energy are interchangeable," Pauli said. "That's what $E=mc^2$ means!"

"Are you saying that the very smallest particles in the universe are made up of pure energy?" Jung asked.

"Precisely."

"But in the very first lecture you told me to split everything into two groups, the stuff and the forces. Now you are telling me there is no stuff at all?"

"Well, no." Pauli reasoned. "Just because everything originates from pure energy does not mean there is no stuff. Don't get so lost in the abstract that you disregard the reality right before your eyes." He said wisely.

"It seems that every time we discuss a new concept, I am left with more questions than when I started." Jung relented.

"In science, when every answer creates more questions, you must be on the right track!" Pauli declared. "Alright, let me introduce you to another concept. Before a particle exists as a stable packet of energy it is popping in and out of existence."

"What does that mean?" Was all Jung could say as the gravity of Pauli's words sunk in.

"A particle becomes matter after waveform collapse, right?"

"Right."

"Wrong." Pauli stated defiantly. "Even as a waveform that packet of energy can be stable."

"Okay." Jung shook his head not knowing what to think next.

"Stable packets of energy have achieved the particle's ground state." Pauli explained.

"A ground state?" Jung asked.

"Yes, a ground state is like being at an equilibrium. It's when a particle exists in a state where it's using the least amount of energy to exist. The same concept is demonstrated in the path of least resistance." Pauli said.

"Ah yes, electricity follows the path of least resistance; that's why it flows down wires but not through insulation."

"Precisely. The same idea applies to chemistry and physics. Do you remember what an atom is made of?"

"Yes." Jung said, trying to follow what Pauli was saying even though it felt like he already knew that part. "Atoms are made of protons and neutrons in the center with electrons in orbit."

"You asked why the electrons are magnetically held in orbit, not touching the protons like we would expect when two magnets stick together."

"Yes, I remember," Jung said impatiently.

"Well, this is the perfect example of something in its ground state. The electron uses less energy when it is in an atom's orbital ring than if it was shoved right up against a proton. Each particle naturally tries to exist in whatever state uses the least energy. In this case a stable particle is just a packet of energy using the least amount of energy it can. Its ground state."

"Okay, then what is happening before it is in its ground state?"

"It is an unstable particle," Pauli explained. "All you need to understand is that particles can pop in and out of existence until they reach their ground state. When a particle pops out of existence it did not have enough energy to be stabilized. Once a particle reaches its ground state, it can no longer pop out of existence."

"The particle is trapped in reality." Jung said.

"That's right, until something else interacts with it. In order for a particle in its ground state to change it requires more energy from somewhere."

"So, before particles become stable packets of energy they might just pop out of existence because they don't have enough energy to reach their ground state. Wow." Jung said with a long sigh. "So, all of our human reality is just energy that might have popped out of existence but for some reason stabilized."

"It has become apparent in modern physics that reality is not as real as we once thought, and yet it is as real as you are." Pauli added.

"I think, therefore I am." Jung said, thinking very deeply after going down a mental rabbit hole in which he started to forget he was real altogether. "So that's it then? The smallest level of existence we know of is just pure energy; but how, or perhaps the better question is, why does the energy become matter at all?" Jung asked.

"HA! Always pushing the limits! I knew I liked you Herr Jung! To answer those questions, you must understand Quantum Field Theory, or QFT for short. QFT is the combination of special relativity with quantum mechanics. Hmm," Pauli thought aloud, "how to explain this."

"The name makes me think of that electromagnetic field you described before." Jung connected the name with the idea of a field of energy.

"Yes, that is a good place to start. First you need to understand the idea of fields." Pauli reasoned. "The EM field exists all around us and excitations of energy pass through it. Actually, the energy doesn't pass through it; the field obtains the energy."

"Where is the quantum field?"

"The fields exist all around us and even within us."

"So, we exist in quantum fields?" Jung asked.

"Yes, everything exists within the overlapping quantum fields throughout all of reality. Let me say, do not confuse these quantum fields with a field like the EM field. The EM field propagates the force."

"Meaning?"

"Meaning the EM field is only active when there is EM energy or electric charge to interact with it." Pauli said. "The quantum fields are the very fabric of our reality." He was distracted by his stomach rumbling.

"This is the conversation I have been hoping for!" Jung said, leaning forward from the edge of his seat.

"Is that the time!?" Pauli exclaimed after glancing at his watch. "I hate to do this to you, but I am afraid I must be going. I have an appointment with Dr. Einstein to review some of his work."

"Say it isn't so." Jung said. "I suppose I can wait another week to find out what the fabric of reality is. Until then I will just have to let my imagination run wild." Jung said before the two said their weekly goodbye. [FACT 79]

FACTS

76.   Is our entire physical reality being constantly generated in every passing moment? Possibly. This concept will be explored in Chapter 27 and Fact 86.

77.   Detecting particles does change them but the technology has improved significantly. We no longer need to totally eradicate a particle to detect it. Instead, we can deliberately shoot other particles, such as photons, at the particle-we-want-to-detect, then we detect those photons. Yes, this sounds like the same description of how we see things. The difference is we are deliberately shooting the photons in a predictable way allowing us to calculate the difference in expected value (if there was no particle) as compared to the observed value (based on how the particle-we-want-to-observed changed the photon we shot.) Even this indirect means of observing particles causes waveform collapse due to the interaction.

The inherent flaw described in the narrative is accurately described and has led people to believe that human consciousness is the direct cause of waveform collapse. Waking human consciousness does not cause waveform collapse. [Ref 68]

78.   Pauli and Jung would have only been familiar with the Copenhagen interpretation at the time this narrative portrays.

Copenhagen interpretation – Developed as a series of theories beginning in the 1920s with the primary contributors being Niels Bohr, Werner Heisenberg, and Max Born. The Copenhagen interpretation refers to many theories but for the conversation on determinism, it postulates that true randomness does exist in quantum mechanics and therefore the universe is not chained to a single destiny.

The Bohm-Interpretation, or the de Broglie–Bohm Theory – First conceptualized in the 1920s, Louis de Broglie abandoned it before it was reinvigorated by David Bohm in 1952. This theory postulates there are unseen waves which act as guides for energy as it localizes into standard particles. This means that what happens in objective reality is decided before it happens.

The Many Worlds Interpretation – Postulated in 1957 by Hugh Everett. This interpretation removes the need to ask the question if the universe is predetermined or not, in this case it is truly infinite and therefore every possible outcome physically exists. Instead of predetermination versus free will, we just don't know if we are in the universe where the particle in question is spin up or spin down. This interpretation requires an entire universe to exist for every single particle in superposition. [Ref 69]

These interpretations have led to different conclusions based on the individual researchers' preferences. If you see headlines suggesting something happened in other dimensions, the researchers involved likely assume the Many Worlds Interpretation is accurate. The articles do not usually explain which theory the researchers subscribe to, which can cause confusion for people unfamiliar with the fact that mainstream physics does not have one clear interpretation that all the experts agree on.

79.   Quantum Field Theory is not Quantum Mechanics. Quantum mechanics is relativistic, meaning solutions change as things get close to the speed of light or have extreme amounts of mass. QFT combines Relativistic Quantum Mechanics with Special Relativity and has been proven as the most

accurate theory in all of science. (Assuming the researchers had valid data entry practices.)

QFT follows relativistic invariance meaning that at those extreme speeds and masses the solutions are constant regardless of the [Special] relativity.

Chapter 26

The Boson of Creation

"I believe we are making progress in your therapy. You have come a long way." Jung said to his, now, good friend Wolfgang Pauli. The pair had just worked through a rather emotional first-half of Pauli's weekly therapy, which had been quite fruitful.

"All thanks to your methods." Pauli said, sitting in the same chair where he started his counseling months ago. Nowadays the chair felt so comfortable it was like a second home.

"Last we spoke you told me about the dream with the dragonfly; are you still having it?"

"Yes. Though it was different this time. As much as I didn't want to admit it, I think you may have been right about those symbols," Pauli relented. "Since I acknowledged the teacher represents the work I was distracting myself with, the dream changed. I seem to have graduated to the next level of the dream's message."

"For deeper insight, perhaps." Jung added.

"I equate it to a series of lectures," Pauli said. "The person giving the lecture has one overall message but must provide an understanding of more basic principles before finally describing the thing they wanted to describe all along."

"What was different about the dream this time?" Jung asked.

Pauli's voice adopted a more curious tone. "The dream starts the same, in the school with the breeze but the dragonfly is different. It starts big and red but dissipates. It shrinks as it flies at me, and actually, it turns from red to a cool blue."

"Very interesting." Jung said, leaning forward.

"The teacher is not angry with me anymore. Instead, he gestures for the dragonfly to land on his desk, which it does."

"What do you think that represents?" Jung asked.

"I believe it is all a self-portrait of me making peace with my anger. I've done a lot of thinking, and I have begun to forgive myself for the selfishness that led to my failed marriage."

"That is a big step," Jung encouraged. "Now to the second part of the dream. What does a goose represent to you?"

"I don't know." Pauli said bleakly.

"Actually, before we get into that, I have done some research on dragonflies," Jung said. "Did you know they change color throughout their lives? Also, they can exist for years under water as larvae before they go through a metamorphosis, just like butterflies do, which gives them the form we are familiar with."

"That is interesting but what does that have to do with this? You think the changing color means something?" Pauli said.

"Well of course. Your mind chose to show you the dragonfly as a different color because it now represents something different. But in the second part of the dream when it charges you from across the lake, is it blue now as well?"

"No, it is still big, red, and angry."

"I thought that might be the case," Jung said. "Is it possible the dragonfly is changing color in the first part of the dream but not the second to highlight the fact you still have something else to acknowledge?"

"Like what?" Pauli grumbled.

"We must figure out what that goose represents." Jung said with just a little too much enthusiasm. Pauli noticed and glared at Jung, realizing he already figured out what it all meant.

"Out with it then!" Pauli said, feeling both amused and nervous.

"Okay," Jung relented. "Have you ever heard of the mother goose stories?"

The smile on Pauli's face was immediately replaced by a frown and an accompanying slouch.

Seeing the change in body language, Jung immediately adopted a somber tone. "Is it possible that you have not allowed yourself to express how you truly feel towards your mother?"

Pauli had already been emotional in today's session. Now he let out a deep sigh and put his hands up, covering his face.

"She took herself away from you," Jung said, "and if that makes you feel angry, that is okay." Pauli spiraled into his emotions like a satellite off orbit crashing to Earth.

"Damnit, why did she do that!?" Pauli shouted as he hit his chair with a clenched fist. "I relied on her. She helped guide me through life, and she abandoned me!"

"Let it out my friend."

"And what kind of person am I that I actually feel anger towards someone because they died?" Pauli sobbed.

"It is ok to be angry; *feel your feelings*." Jung said softly as tears began to stream down Pauli's face.

Pauli had a good cry and finally acknowledged the anger he harbored towards his mother because of her suicide. Up until that point he had been burying his emotions in his work and blamed all his dreams on his failed marriage. To help him process all the implications he paced around the room and even stopped to look at Jung's window which had been boarded up from the previous week's Pauli Effect.

Pauli reminisced on some old memories of happy interactions with his mother before he was able to return to the emotions at hand and learn from them.

"I told myself I already addressed everything that was bothering me but the truth is, I've still been distracting myself with work." Pauli figured.

"Yes. I believe the dragonfly did not represent your anger, instead it represented how your anger was negatively affecting you. Your teacher represented the work you were distracting yourself with. In the next scene it showed how your anger towards your mother was eating away at you."

"I am still amazed that my own mind attacked me by siccing a dragonfly on me in a dream."

"When you attacked yourself in the dream it was just symbolism for how your emotions were tormenting you in real life and only your unconscious mind saw what you were putting yourself through." Jung suggested.

"Right." Pauli said as he took a deep breath to process everything he was going through.

"In time, as you acknowledge and address your anger, you will begin to forgive your mother." Jung explained.

"I suppose this is all part of the process you told me about last week. Was it, individualizing?"

"Individuation." Jung gently corrected, "and yes, the first step, in this case, would be to acknowledge the aspects of yourself which do not match the ideal version of who you want to be."

"I felt guilty because I was angry at my mother for dying, I suppose."

"And the guilt made you suppress the anger because you think of yourself as a good son, but according to your own definition, a 'good' son wouldn't be angry." Jung connected Pauli's archetypal self and persona to his actual feelings as the timer rang out, ending another week's counseling.

Pauli drew a long deliberate breath. "And now for the part I know you have been waiting for; let's discuss the quantum fields!" Pauli said as lively as he could, though pain still lingered in his voice.

"Herr Pauli, I'm not sure we should jump right into this lesson. You just had a very deep realization and I don't want you to-"

"Nonsense." Pauli interrupted, sounding much more genuine after another deep breath. "I will not put on an act and pretend I am fine but lecturing you in physics has become a therapy in and of itself."

"I am honored you feel that way." Jung said sincerely, pausing a few moments before saying, "I was thinking about our conversations last week but I was having a hard time picturing what a photon looks like."

"What does a photon look like?" Pauli said thoughtfully. "That's a nonsense question; it looks like light."

"I mean while it's a waveform bouncing through both those slits at the same time."

The Boson of Creation

"It is still just light." Pauli reiterated.

"But how can I conceptualize it?" Jung prodded. "Would it literally look like one of those waves you drew?"

"You are combining different aspects into one." Pauli said. "We literally see photons every day. Its light; don't overthink it. One photon by itself is still light."

"Well, what about a photon with a frequency outside visible light?" Jung asked.

"It isn't so simple," Pauli explained, "and what it looks like to humans doesn't matter for physics. The waves I drew are a measurement of a photon's energy, not what a photon would actually look like."

"But why is it so hard to imagine?" Jung pressed.

"We have to break down each property," Pauli explained, "define what each would look like then leave you to guess how they might interact as we piece it all together."

"What do you mean break down each property?" Jung asked.

"Photons have a spin, an electro-chromodynamic color, a velocity, a frequency, an amplitude and other properties that all make them unique. Each property might make them 'look' different but it's all just light to us." Pauli ranted. "Think about your shoes."

"My shoes," Jung said looking at his feet.

"One of their properties is color; black, brown, white, grey, et cetera. Another property is luster; shiny, matte, glossy." Pauli explained. "Glossy black shoes would look quite different than matte grey ones even though they are both shoes. We would need to know what each property looked like individually, then figure out if combining different properties changes the look of others but here's the thing; these properties are literally too small to look like anything."

"How is that?"

"Color is what we see when EM energy is a certain frequency. The differing sizes of the physical wavelength give us colors. We are discussing properties that are smaller than one of those wavelengths, therefore there is literally no color there." Pauli explained.

"Smaller than color." Jung thought aloud.

"Furthermore, it is possible that photons have properties that exist in higher spatial dimensions which would literally be impossible to imagine."

"In other words," Jung confirmed, "we can only understand what a human can see and a photon is the limit of what we can see so if we break it down and examine its parts, we literally cannot see it. Therefore, what a particle looks like doesn't make any difference to anything."

"Yes!" Pauli exclaimed. "What we need to focus on are the properties that make the photon what it is and act how it does. And for you to understand that we should discuss the quantum fields." Pauli sounded tired but determined to complete this lecture.

"If I may ask," Jung said cautiously, "well, it's just that, as we began the previous lectures, it seemed like we were discussing the absolute final explanation, yet every time there has been some smaller scale to explain the origin of the previous topic."

"Yes…" Pauli said, nodding.

"Well, how many more layers of reality are there? I don't mean to say I'm not enjoying our discussions, it's just that I am trying to hold a complete picture in my mind and I am starting to believe physics has no end at all!"

"Ha! I will offer two answers," Pauli said defiantly. "The first is that I love the sentiment you just understood; physics will never end. As soon as we answer one question a dozen more questions arise. But here is my second answer; Quantum Field Theory will be my final lecture to you. Once I explain what we think we know of the quantum fields I will be left with nothing more to provide besides questions. Questions that no one has answers to."

"*Yet.*" Jung said enthusiastically.

"That's the spirit!" Pauli spat with a hearty chuckle.

Was that laugh really genuine? Jung asked himself, or is he masking his pain again. He hasn't had time to deal with it and now we are focused on something totally different.

"Let's get right to it." Pauli continued before Jung could say anything. "Remember the different perspectives we can view reality from? The cosmic perspective shows us the universe from the largest celestial bodies."

Jung wanted to follow the lecture, so he closed his eyes and pictured reality at the different scales as Pauli mentioned each one.

"There are many microscopic perspectives; cellular, molecular, and atomic. Still going smaller we get to the quantum perspective where things get strange."

"Strange indeed." Jung added.

"From the perspective of human reality, I want you to imagine an invisible field all around us. It is not bound to the laws of physics that you and I are, and yet it permeates throughout the entire universe. This field is literally creating all of our physical reality by focusing pure energy."

"Right, $E=mc^2$ so all physical matter is actually just energy within a quantum field." Jung said, remembering the previous lecture.

"In Quantum Field Theory, or QFT, there are many of these different fields."

"Next to each other?" Jung asked.

"Well no, all the fields exist throughout the entire universe, overlapping."

"How many fields are there?"

"There are different theories but our best guess is one field for each type of fundamental particle." Pauli said.

"So, three? One for protons, one for neutrons and one for electrons?"

"Now I see where I've lost you," Pauli said. "I have not adequately discussed the fundamental particles."

"That's right." Jung remembered making the same mistake last week.

Pauli smiled playfully at Jung's frustration. "Protons and neutrons are not fundamental particles, but photons and electrons are. There are other fundamental particles I have not covered."

"Okay, I think I understand." Jung said. "We are still at the quantum scale; smaller than protons and neutrons but larger than, well, raw energy."

"Right. The fundamental particles can be grouped into three types: quarks, leptons and bosons. Let me draw it out for you." Pauli said as he scribbled away. "But let us not get confused. At this point I still want you to conceptualize everything as either forces or matter." He drew out all the fundamental particles grouped together.

Standard Model of Elementary Particles

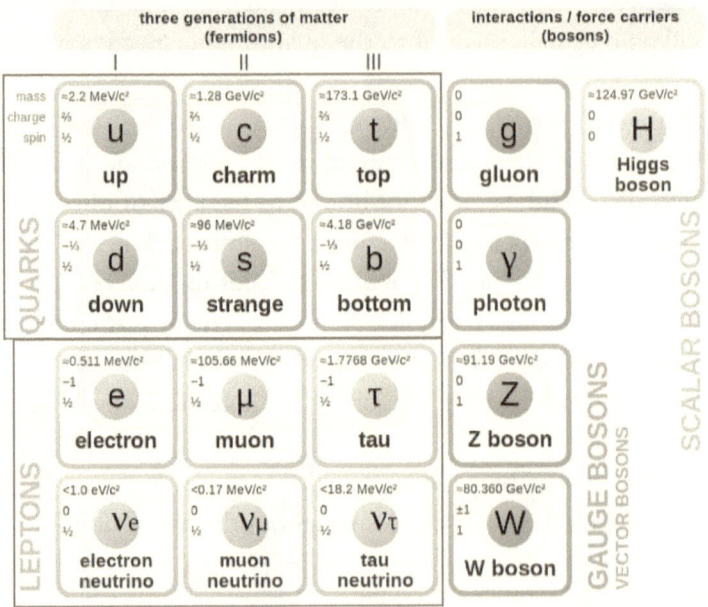

"Atoms belong to the periodic table," Jung said, "but particles belong to this?"

"*Fundamental* particles, yes! Think of this as the periodic table for quantum particles, we call it the Standard Model. There are 12 fundamental particles that fall into the category of the physical matter of the universe. However, most of these particles are not stable and therefore do not provide any direct input to our human reality, or this conversation, which is why I only told you about electrons and up and down quarks. For QFT it's good to know there are six quarks and six leptons." Pauli explained.

"If only three are the stuff of reality, then what are the other nine?" Jung asked.

"To be clear, out of those 12, all of them have mass and therefore are matter. Only the up quark, down quark, and the electron directly build to atoms and therefore everything we see in the human reality." Pauli explained.
[FACT 80]

The Boson of Creation

"I see." Jung said with a pause. "What about these things you labeled as bosons?"

"Those are the particles that carry the four forces. The gluon carries the strong force, the photon is electromagnetism and the z and w bosons both mediate the weak force."

"That fifth boson, there?" Jung asked pointing to the last one all the way on the right. "That must carry gravity, the only force you didn't mention."

"Wrong again, Herr Jung." Pauli said, enjoying his friend's frustration. "The graviton might be another force-carrying particle, but we are yet to discover it. It might not exist at all, so I did not include it in this chart. This last one is the Higgs Boson which is very special. It is otherwise referred to as the 'God particle,' which literally creates all mass. But first you should understand these fields I mentioned." Pauli said.

"Now wait a minute! You cannot introduce something called the 'God Particle' then not explain it." Jung said vehemently.

"It's one misnomer after another!" Pauli spat waving a hand angrily, "The short of it is, all particles, even photons interact with the Higgs boson. However, you are curious about the Higgs mechanism not the Higgs boson. All particles interact with the Higgs but don't necessarily gain mass. The Higgs mechanism is specifically the gaining of mass either by spontaneous symmetry breaking down or, for fermions, Yukawa coupling."

Pauli watched as Jung's eyes glazed over in real time.

"There is no 'God' in the equation just as there is nothing spontaneous," Pauli ranted. "The Higgs mechanism is what causes particles to gain mass."

"Mass," Jung said thoughtfully, "you mean the Higgs particle is what creates all the physical stuff in the universe?"

"Specifically, mass. The Higgs adds mass to raw energy, effectively creating the physical world and everything physically in it."

"How?" Said Jung.

"Think of how charged particles interact with a magnetic field. If the particle has an electric charge, then it naturally interacts with magnetic fields. Neutrons can exist in the same magnetic field but have no interaction whatsoever. Well, the photon is like the neutral particle in the example."

"It interacts because it is in the field but has no affect." Jung confirmed.

"All the particles that have mass would be like electrically charged particles. When these particles interact with the Higgs it triggers the Higgs mechanism just as a charged particle requires a reaction in a magnetic field."

"The God particle." Jung said in awe. "So absolutely everything that exists has interacted with the Higgs Boson. And you have observed this? It is proven?"

"Indeed, it is, however I oversimplified my explanation," Pauli answered. "It's worth mentioning that the fundamental particles that carry forces also have mass and therefore even the forces themselves rely on the Higgs field; all the particles, except the gluon and the photon that is." [FACT 81]

Jung did not notice that Pauli referred to the Higgs as a field rather than a particle.

"One fundamental particle at the very foundation of it all." Jung stated ruffling his brow. "So, you have found God."

"The name would have you believe that but let me challenge that thought." Pauli said. "All we have done is define some physical process that naturally occurs in the universe. In what way is that God?"

"According to the Bible, God is not some man in the clouds. He is unfathomable. Isaiah 40:28 says, 'God is without beginning and without end and in this way is unsearchable.' Yet again, Herr Pauli, I find myself contemplating those same two questions; what is consciousness and what creates reality?"

"This is no Sunday School lecture I've ever heard! In Christianity God is clearly understood as a man, but what does it matter? We are discussing physics!" Pauli spat.

"Before the many translations over thousands of years diluted the original, the common Christian understanding was that God was both masculine and feminine and what's more is that He is incomprehensible. What if God is the very process that your physics describes?"

"What if God is a goat or a man or the whole damned universe?" Pauli retorted, "it has nothing to do with what we are discussing."

"Actually, that's a good point," Jung said disregarding Pauli's pleas to stay in the confines of science. "If God is this process of creation, then would He not also be the things being created?"

Pauli folded his arms and rolled his eyes.

"What if our individual consciousness is somehow connected to God and it is He, *through us*, that influences which properties particles obtain when they collapse out of superposition!" Jung said, sharing his most vulnerable idea.

"The term God is too broad, or unknowable, as you mentioned. The name doesn't belong in physics." Pauli said stubbornly.

"That's a funny thing to say since it was you who mentioned God in the first place." Jung said with a smirk.

"A fair point but I didn't name the particle!" Pauli said quickly.

Jung wasn't sure why the idea of God was such a touchy subject to his friend. Perhaps, Jung pondered, his many years of interacting with others in academia has caused the subject to be a mark of ignorance and therefore carry negative connotations.

After an awkward silence Jung spoke first. "I imagine the God particle's name is derived from its function to create physical stuff, which would be a simple association to the concept of God via creation." Jung said calmly. "To get back to QFT, should I ask how the fields interact or would it be better to ask how each individual field creates each individual particle?"

"Reaching to God as an answer in physics is to forsake the process of the scientific method." Pauli said, seemingly confirming Jung's guess, but accidently steering the conversation back to religion. [FACT 82]

"I just wonder, when the observed phenomenon you are trying to explain is the very definition of God then what will scientists do?" Jung challenged.

"Luckily for my argument, we have no definition of God that any two people can agree upon." Pauli said cooly. "I think we should start by discussing just one quantum field and the energy it hosts, then we can discuss how the fields interact."

"Very good." Jung said shifting his thoughts back to the invisible fields creating reality.

"A quantum is the smallest bit of energy that can exist. By definition it must be a single quantum particle. In Quantum Field Theory we are going to talk about subquantum energy as well."

"*Sub*-quantum meaning smaller than quantum? But you just said that a quantum is the smallest bit that can exist." Jung said, puzzled.

"That is exactly correct, Dr. Jung. Subquantum energy *doesn't exist.*" Pauli said with a sly grin.

"Is this what you were referring to when you said rogue electrons can pop in and out of existence?"

"This is exactly what I was referring to," Pauli said. "Before a fundamental particle has enough energy to exist, it is just an excitation of energy in one of the quantum fields." Pauli explained. "It's called subquantum kinetics." [FACT 83]

"And in this field, it is just pure energy?" Jung asked.

"Yes. In a way, the field is the mechanism that focuses the energy. When there is energy present, you see."

"But I don't understand," Jung interrupted. "How could this energy not exist? Clearly, it exists somewhere."

"To your point, it does exist in the sense that it is more than nothing. The subquantum energy I am speaking of is called virtual. Virtual particles and the quantum fields themselves are the bridge from our real universe to whatever ether exists on the other side of creation." Pauli tried to explain.

"Are you saying the virtual particles exist in some other reality?" Jung asked.

"No, no, no, I should not have said it like that. They do not have enough energy to be a single quantum. Though the really interesting thing is they can influence actual quantum particles." Pauli explained.

"Let me try to understand what you are saying because it seems like you are describing three totally different realities." Jung attempted to summarize with an analogy. "The first is our human reality from the subatomic to cosmic scales but then there is another place, seemingly opposite of that, which is totally out of our existence. It is the source-energy of our reality where virtual particles exist. Finally, there is another reality, the quantum fields themselves perhaps, which seem to bridge the pre-creation, energy-reality to our human-reality by physically creating it at the quantum level."

"I do not care for this analogy." Pauli said stubbornly. "I have certainly never conceptualized it this way. It is all part of one reality but perhaps this analogy will help you begin to understand it all, then I can correct your understanding later."

"I am guessing there are still many unanswered questions regarding this subquantum world." Jung stated, thinking about taking another leap of faith to explore a deeper concept.

"I think it would be more accurate to say there are unrealized implications." Pauli explained. "The math tells us what we need to know to understand the basics of Quantum Field Theory and how everything is produced by the fields. Let's define and deconstruct a single quantum and how it exists in the quantum field, then we can explore subquantum energy." [FACT 84]

FACTS

80.   You previously learned that a proton is two up quarks and one down quark. The truth is all the quarks and leptons directly contribute to our physical reality. Many quarks and leptons are constantly flowing in and out of existence within each proton or neutron, directly contributing to each particles' existence and potentially their specific properties. The thing is all those extra particles are constantly cancelling out to maintain the balance of two up and one down quark (in a proton). This will be explored in History, Repeating Itself Part II; *Treasure Lies Within*.

81.   The Higgs boson is a real particle that induces the Higgs mechanism. This is a proven phenomenon and all mass in existence has interacted with the so-called "God Particle." Reality then, might be process driven. Energy flowing through the Higgs mechanism literally creates our reality. [Ref 70]

Pauli and Jung would not have known of the Higgs Boson. It was theorized by Higgs, Englert and others in 1964 and only discovered in 2012.

82.   Even today many people in science view anything relating to "God" as scientific blasphemy. If they had the slightest shred of creativity, they would recognize that mysteries attributed to God are nothing shy of the greatest opportunity in science. If you simply disregard these mysteries, you must not be up to the challenge.

83.   Subquantum kinetics is an emerging, but very real, field of study that describes virtual particles and the energy that comes from somewhere, somehow interacting with the quantum fields, perhaps even in subquantum fields.   [Ref 71]

84.   Quantum fields, virtual particles, and randomized outcomes (of particles upon waveform collapse) are all accurately depicted in the narrative.

Chapter 27

Foundational Fabric
and the Particles from Nowhere

"Very well." Jung continued the conversation. "What is a single quantum?"

"Quantum particles are discretized."

"What does that mean?" Jung blurted out.

"Each particle contains a certain amount of energy. Think of counting."

"Counting?"

"Yes. What comes after 1?"

"Two?" Jung shook his head.

"What about 1.1? 1.2? 1 and a half?"

"Well, I don't know!" Said Jung almost laughing at the ridiculous line of questions.

"For quantum particles there is no 1.1 or one and a half. There is only one whole packet or two whole packets, or three, et cetera."

"Is that what discrete means?"

"Exactly," said Pauli. "To understand quanta though, you must understand the field that is generating them." Pauli lectured. "Each type of fundamental particle can only be generated from its own specific quantum field. Meaning there is one infinite field for each type of quantum particle. The fields all overlap throughout all of reality but probably don't exist in any physical way."

"Wait," Jung interjected, now confused with the various terminology, "you just said quantum particles but before you said fundamental particles. What is the difference?"

"They are the same thing." Pauli said, amused that his friend was struggling but also impressed that he recognized minute differences in the terminologies.

"Okay, got it." Jung said. "The quantum fields go out infinitely in every direction but what is the field itself?"

"Yes, allow me to describe a field by itself. No particles or subquantum energy, just the field. First and foremost, the quantum fields are the foundation of reality."

"Whoa. But I thought space-time was the fabric of reality." Jung stated.

"Yes," Pauli said, "space-time is a layer of the fabric of reality but so are the quantum fields."

"Wait. Are the quantum fields the same thing as space-time?" Jung wondered.

"If they are we certainly don't think about it that way," Pauli explained. "I mean they don't really relate to each other. Space-time is how we describe our four-dimensional reality from our perspective but QFT explains, or views, reality from the perspective of quantum particles."

"So, they are the same thing." Jung stated for confirmation.

"It's a moot point!" Pauli challenged. "They are the same thing just as fundamental particles and forces are the same."

Jung was a little confused. "But those have different functions."

"And yet they are all quantum particles." Pauli explained. "The point is we are talking about totally different things that may have commonalities; but what they have in common just doesn't matter."

"Meaning?" Jung asked.

"Are space-time and the quantum fields the same thing?" Pauli asked rhetorically, "sure, why not? My point is you are oversimplifying something for no reason."

"Oh," said Jung, "I am trying to conceptualize a single fabric of reality but reality is more complicated than that."

"One might even call it, unfathomable." Pauli said, mocking Jung's earlier biblical reference. Jung leaned back in his chair and chuckled.

"Before we get any further off track let me describe a quantum field. Each field has a minimum amount of energy; one particle's amount."

"Which particle?" Jung interrupted.

"Every fundamental particle has its own corresponding field that generates it." Pauli said.

"So, the up-quark field has a minimum energy, equivalent to one up quark." Jung understood.

"Precisely," Pauli continued, "the field can host energy but it is not the energy itself." Pauli explained.

"But you said the field is not physical. So, the fields don't actually exist? I still don't understand what the field is."

"Potential." Pauli said definitively. [FACT 85]

"Potential?" Jung asked, hoping for a less abstract explanation.

"The actual field does not have anything tangible in our reality that you need to conceptualize besides a minimum amount of energy equal to one particle. Instead think only of the potential it facilitates which becomes real when energy flows through."

"Oh," said Jung, "the fields *become* real when energy *flows* into them." He repeated.

"Yes, and when enough energy is focused in, let's say, the quark field a quark packet might interact with the Higgs, forcing it to gain mass and pop into existence. Then if something in our reality affects that quark's energy, it can simply leave the quark quantum field and enter, let's say, the photon quantum field causing the energy to be expressed as a photon in human reality. I am oversimplifying it but the energy itself exists in and outside of our reality." Pauli said, sitting back into his chair, looking accomplished with his lecture.

"It's outside our reality." Jung repeated in awe. "What does that mean? I'm afraid I still don't understand what a quantum field is."

"Think about light shining through a stained-glass window." Pauli said. "The stained glass of the window is the quantum field; the light is the primordial energy and when the light goes through the stained glass it can only shine through as the color of the glass."

"Okay, I understand that," Jung said.

"Picture it in your mind," Pauli instructed. "The stained glass is the quantum field."

"It's colorful." Jung replied, imagining the inside of a church.

"Now take away all the light from the scene in your mind."

"Okay." Jung watched the inside of the church get dark.

"What do you see?" Pauli asked for confirmation.

"A dark church with dark windows."

"No!" Pauli said definitively. "You are in a room that light can only enter if it goes through the stained-glass window. You are in total darkness and you cannot see anything because there is currently no light shining through."

"In that case, all I see is darkness." Jung relented.

"If you see any light at all it will be the color of the glass it is shining through," Pauli encouraged. "You know the glass is there, but you can't see it until light starts shining through."

"Okay," Jung said. His imaginary church lit up with the various colors from the stained glass.

"Each color is a different quantum field and yet they are acting as one to show you the picture on the glass." Pauli said. "Now imagine the stained glass is not a church scene. Instead, it is just two large panes of glass; one red and one blue."

"The red one might be up quarks and the blue one could be down quarks?" Jung guessed.

"Sure," Pauli said, "and in your room you can see some red light and some blue light but where the light overlaps you see-"

"Purple!" Jung blurted out.

"Which, in this analogy," Pauli continued, "would be how a proton or neutron is formed by up and down quarks via the quantum fields."

"I get it now!" Jung exclaimed.

"Good, however, let me correct that example." Pauli said.

"Here we go," Jung said, rolling his eyes at his short-lived understanding.

"That was a two-dimensional example of a three-dimensional thing." Pauli lectured. "If the stained glass is the quantum field, then it does not only exist as a single pane in a wall. The light source is not on the outside of it and the colored light it produces is not inside a room. Every color of glass exists in all directions throughout every point in space. If you see purple light, it wasn't because something somewhere else shone through; the purple light you see is directly within the quantum field in the location you see it."

"Okay," Jung struggled to understand. He imagined being in the church seeing purple light hitting the ground and said, "the purple light is not on the ground, the glass producing the color exists in the exact place I am seeing it." Jung imagined the red and blue glass windows growing thick, expanding outwards until they overlapped throughout all of reality. The different colors of glass turned clear where they were not hosting energy and the purple light was no longer on the floor. He looked up to the window where the energy actually was. The blue and red stained glass was now purple because the two quantum fields were both hosting energy in the exact same place, creating a purple light.

"It is two fields overlapping," Pauli continued. "Both fields are hosting the light or the source energy individually, which follow some process, interacting with the Higgs field becoming physically real and combining in physical reality to create a proton or neutron!" [FACT 86]

"I see!" Jung exclaimed. "The field itself is like a bridge because it hosts the energy while itself not needing to exist at all. I guess another way to say it is the quantum field only physically exists when energy is present, and the field only exists as the particles it manifests." He theorized.

Pauli thought for a moment about this concept. "That is an interesting way to describe it. Every fundamental particle is an extension of its field which means fields do physically exist as the particles themselves." Pauli said pensively.

"The energy is focused by the quantum fields bursting into our three-dimensional reality, seemingly without warning and at random, which is how real matter becomes…" For a moment Jung trailed off as he tried to find the correct word to finish his thought until he realized he already said it. "That's how real matter becomes." He said definitively.

"Once the quantum field has focused enough energy to equal one whole particle packet," Pauli explained, "the real particle, physically in our reality, either as pure energy or a particle with mass, will either decay almost immediately because it is not stable, or it will stabilize and be trapped here in our reality until something else interacts with it."

Jung asked, "What causes a particle to stabilize?"

"Energy in the quantum field can be 'tuned' by surrounding particles for stability."

"What about that God particle? Didn't you say the energy must interact with it?" Jung added.

"Yes, yes, the energy interacts with the Higgs particle via the Higgs quantum field." Pauli said.

"But what causes that interaction?" Jung asked.

"I believe it is a naturally occurring process. There must be some law of physics that has not yet been discovered which requires the focused energy to interact with the Higgs field forcing it to become physically real, but this is just my guess."

"So, it's just a law of physics like how the law of gravity forces an apple to fall if its stem brakes and there is nothing else to hold it to its tree." Jung understood. "But when does the energy interact with the Higgs; before or after it stabilizes as a whole packet?"

"Let me ask you a question; when does your brain apply the heuristic about the light from above us, is it before or after it has analyzed the shapes and colors?"

"Well, we don't know! It could be either or it could even happen at the same time." Jung explained.

"I have the same answer for you. We do not know if there are steps completed in order or if everything happens all at once," Pauli explained before driving the conversation forward. "Let's conceptualize the quantum fields by talking about how a photon is generated by the sun."

"Okay."

"In the sun's core there are many hydrogen atoms which sometimes collide. When they smash together, it creates a chain reaction that chemically changes hydrogen atoms into helium atoms."

"And I am guessing this is not a simple sharing of electrons." Jung presumed.

"Correct. The particles have a high velocity and smash together which creates a violent atomic reaction that emits electromagnetic radiation and heat."

"Electromagnetic radiation?" Jung thought for a moment on which fundamental particle carries EM energy. "You mean photons; light!"

"Exactly," Pauli said. "Now, what do you think this process looks like in the quantum fields?"

Jung was enthralled to be exploring the very fabric of reality with a real example. He first pictured the sun before his vision magnified to the atomic scale where he watched four hydrogen atoms collide. Inside each atom he saw the proton split from the neutron as heat and a blinding light exploded from it.

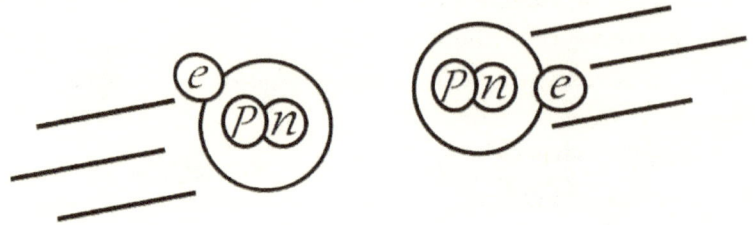

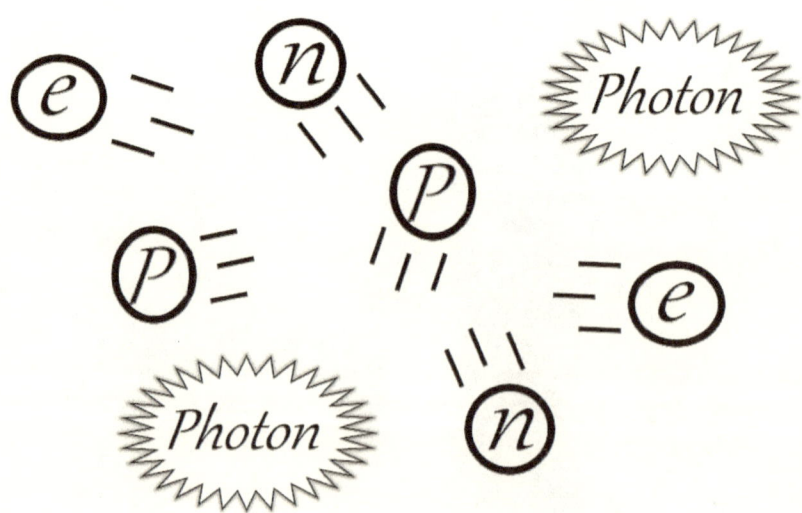

Jung realized he was not at the correct scale. He watched the scene play out again from a lower order of magnitude. He saw inside each proton and neutron. Clumps of up and down quarks were mingled together. The bundles of quantum particles barreled towards each other. As the atoms collided the quarks holding them together split off from one another and, in an instant, some were bound together creating a new atom, helium. The other up and down quarks had somehow turned into photons, which chaotically flew out passed Jung.

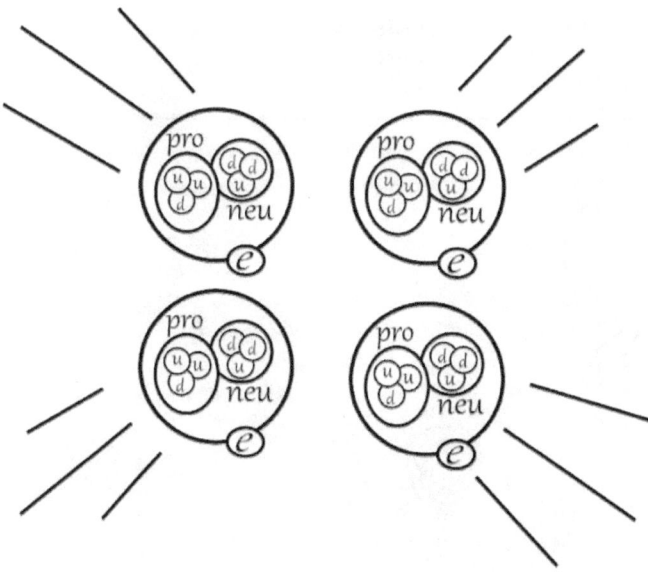

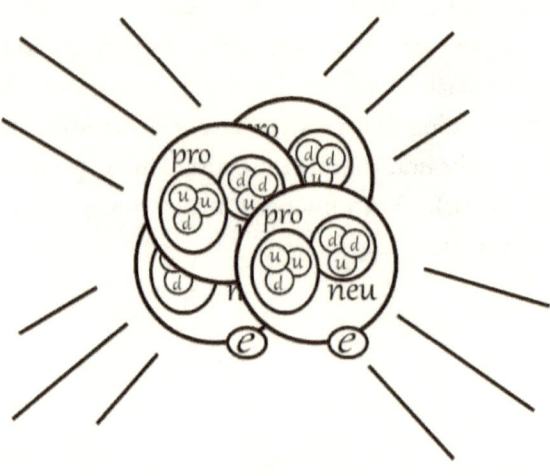

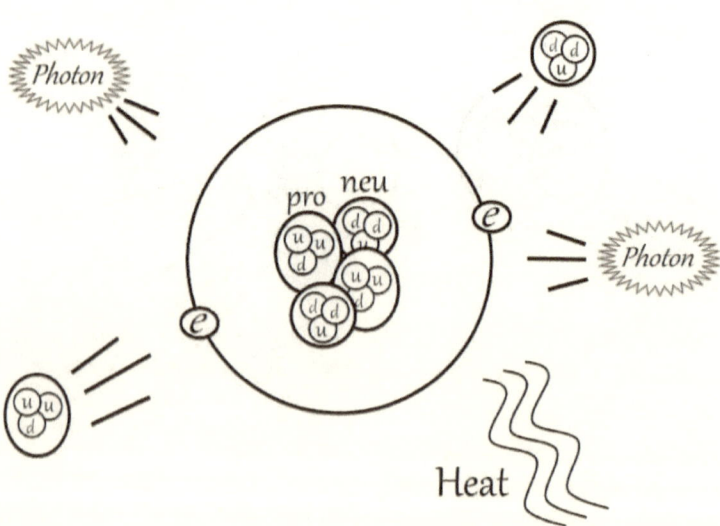

The atoms collided in reality and hydrogen turned into helium and light. But what was happening in the quantum fields that created all this, Dr. Jung wondered. The scene repeated for a third and final time and Jung went into the raw energy of the quantum fields.

He looked around to the different spots in physical reality. From inside the quantum fields, he could see the sun's atoms as if reality was faded to the background beyond a thin white veil. Each up-quark and down-quark was like a three-dimensional ripple of vibrating energy buzzing like bees in the field itself. The atoms smashed together separating the protons and neutrons, scattering the up and down quarks; and in the quantum fields the energy manifesting each particle kept pace with what was happening in human reality.

The energy in the quantum field *is* the particle, Jung reminded himself.

The buzzing and flickering pieces of energy that stayed in the up and down quark fields were suddenly pulled together, as if a rubber band snapped around them, forming the new atom, helium. The rest of the energy disappeared from the quark fields and a new field emerged into Jung's vision. The photon field gracefully received the energy as if it wasn't making any transition at all. Some of the same energy that had been producing a hydrogen atom was now pure light.

"But what about the photons?" Jung said puzzled. "How did the energy in the quark fields transfer to the photon's field?"

"Easily! The energy in the quantum fields can flow from one field to another because quantum fields themselves don't physically exist!" The conversation finally came full circle and ended up outside of our reality altogether.

It was like Jung entered a strange new world where reality was undefine and yet he felt more comfortable there than ever before. "What is the energy that flows through the fields?" [FACT 87]

"Not what, where." Pauli said defiantly.

FACTS

85.   The quantum fields themselves don't physically exist yet are definitely real. Each field is a non-physical medium that hosts energy. The energy is forced to become the particle corresponding to that field. Fields are discretized and always have at least one particle's worth of energy. [Ref 72]

86.   Particles are created by the quantum fields when energy is focused in a similar way to what is described in the narrative. Existing particles tune nearby energy in the quantum field, focusing them to become a certain particle. [Ref 71]

87.   The energy-of-creation, or whatever we are calling it, flows through a quantum field to become physically real. Once it is stable and physically real, the energy is no longer 'flowing' from some other quantum field (or some undiscovered primordial place) to maintain that particle in the eternal present moment. The energy flowed and is now static as the particle, until something else interacts with it. (To be clear, the particle's energy is still being hosted by the quantum fields.)

So, is there some Tao-like energy eternally flowing which sustains the present moment? It depends who you ask. Most physicists today conceptualize a static Hilbert space that reality exists in. (Reality being relativistic spacetime and the quantum fields as described by the entire field of physics.) In many physicists' minds the energy only flows within the quantum fields until a particle is stable in reality, at which point the flowing stops.

This requires time to be conceptualized as we experience it in human reality but, like gravity, time might be very misunderstood. Some Theoretical Physicists describe a different concept of time in which they acknowledge there is no future (because it has not happened yet) and there is no past (because individual events have no past [they are the result of causality based on other individual events]). This is evidence of previous present-moments but there is no reason to believe the past still exists. [Ref 93]

This reasoning allows a Tao-like flowing energy to become the intuitive answer to the ubiquitous question; how?

Chapter 28

King of Kings, Lord of Lords, Field of Fields

"I regard consciousness as fundamental. I regard matter as derivative from consciousness. We cannot get behind consciousness. Everything that we talk about, everything that we regard as existing, postulates consciousness."

-Max Planck, PhD Physics
Father of Quantum Theory

"Okay then where is the energy?" Jung asked desperately trying to understand not only the quantum fields but the energy within them.

"What I described is the limit of what we know. This and any further discussion will be nothing more than guesswork with little to no facts to support it." Pauli stated authoritatively. "You see, our scientific understanding currently ends at the quantum fields as they propagate or focus, or perhaps just imprint packets of energy. Presumably this energy has always existed and instead of being created or destroyed it is simply changing by bouncing between quantum fields."

"But you think there is something else?"

"The Big Bang is confounding to me. I understand it was an extremely hot dense mass of everything but how and why seem illogical. One moment before the bang, when it was all still an almost-singularity, there was energy but no mass. Did the Higgs and other quantum fields exist or not? If not, when did the quantum fields come into existence, and how?"

"Why would they not exist?" Jung asked.

"The discretized nature of the fields requires them to have a minimum amount of energy equal to one full particle of that type. At the Big Bang there could have been energy in fractals smaller than even the smallest single particle, so where did it exist? What medium hosted them?"

"You think there is some other field?" Jung guessed.

"You are finally ready for me to complete my lecture series," Pauli continued. "There is one last understanding you need then you will have been introduced to the entirety of reality as we know it. From there it will be up to you to explore the mysteries on the other side of the veil that separates this reality from the great beyond."

"That was a hearty introduction. I'm sure this will not disappoint." Jung said.

"Each fundamental particle has its own field but there is another field," Pauli said dramatically. "This field is the most basic and fundamental reality we could ever hope to understand. It is not the source energy that flows through the fields, but it is where that energy lives." Pauli explained to a wide-eyed Jung who was on the edge of his seat.

"What is this field?" Jung asked excitedly.

"It is an infinite number of one-dimensional points," Pauli explained. "That means the point is as small as anything could possibly be, orders of magnitude smaller than a quantum."

"Wait, a single quark is not one dimension?"

"No," Pauli huffed. "It is extremely small but it still has length, width, and depth so it is 3-dimensional. Are you familiar with the Planck length?"

"No."

"A Planck length is a theoretical measurement that is the smallest anything in the universe could ever possibly be." Pauli explained.

"Finally," Jung said, "we have reached the very end. The most fundamental building block of reality."

"In this, most fundamental field, each of these single points is touching another one. They go out in every direction creating a field of potential." Pauli explained.

"An infinite sea of one-dimensional points makes up the field that is the very foundation of reality itself." Jung understood.

"Each quantum field can only host the type of fundamental particle associated with that field. In the Planck Field, each point can host the smallest amount of pure energy, not attributed to any type of particle." Pauli explained.

"Remember virtual particles?" Pauli continued, "these excitations are outside of our reality but they have the same amount of energy as one full particle." Pauli explained.

"Then what is the difference between that and a real particle." Jung asked.

"*When* is the difference," Pauli said mysteriously. "The Planck length refers to space but as Einstein taught us space is relative to time so there is also a Planck time."

"The smallest amount of time that can exist?" Jung said trying to figure out what that meant.

"Precisely! Virtual particles cannot be detected because they exist for less than one Planck time."

"What is the difference between a virtual particle and an excitation?" Jung intrigued.

"A virtual particle is being hosted by one of the specific quantum fields and it can influence real particles. An excitation is rogue. It is pure, raw energy that only exists in this Planck Field." [FACT 88]

"I'm not sure I understand. The line of reality and not-reality is very blurry now. At what point did we leave reality?" Jung asked.

"The Planck Field, and the energy that begins to focus in it, is the limit of physical reality." Pauli said. "Anything smaller than one Planck length or one Planck time is outside our reality."

"So," Jung attempted to summarize, "all those different kinds of energies are actually just one source energy that is flowing through the various quantum fields and that's what creates different things in our reality. The energy itself is interchangeable. That means there is only one singular type of energy that follows the laws of the quantum world."

"That is quite possible but…" Pauli paused, leaving Jung suspended in his thought of a singular energy type that was the root of all creation. "No one has a damned clue." [FACT 89]

"Ha!" Jung let out a laugh as he collected his thoughts. He was left with only questions, just as Pauli said would happen. "Okay," he focused his thoughts, "back to what we do know. The excitations in the quantum fields can flow from one quantum field to the next which changes what particle the energy is expressed as in our reality, correct?"

"Correct, but there is more to it." Pauli said.

"What causes the energy to move from one field to the next?" Jung asked.

"There are many factors," Pauli said. "Remember, our space-time and the quantum fields all have laws of physics that particles, and therefore this energy, must follow. The energy exists and is simply following the laws for the container it exists in."

"In this example, the container is our reality?" Jung guessed.

"Yes." Pauli encouraged. "Remember the quantum fields are like light shining through a prism. White light splits into the rainbow, not because it chooses to but because it must follow the laws of physics."

Jung thought aloud, "so there is the fabric of reality which is space-time and the quantum fields and then there is the energy that flows through the quantum fields, which gives us human reality."

"Producing all things in existence," Pauli added.

Jung was determined to understand the laws of the quantum fields. "Can you give an example of a real law this subquantum energy must follow?" He asked, hoping to fully understand what Pauli was saying.

"I'm not sure that I can." Pauli said to Jung's disappointment. "We have some observations, but this field of study is so new we don't yet know what is correlation and what is causation."

Jung thought for a moment about what these inherent laws might be. Then a thought popped into his mind, perhaps if I understand how the energy changes from one field to another I will understand. "Can you give an example of something that can force energy to jump to another field?"

"There are so many examples," Pauli said. "Literally anything; if you turn on a light switch you are forcing energy to jump to the photon field where it becomes light for us to see."

"What?" Jung exclaimed. "We can make the energy jump fields?"

"As far as I can tell, actions in our reality are *required* in order for the energy to change in the quantum fields."

"Perhaps I misunderstood your previous comments. It is just like the internal and external worlds of the human psyche." Jung thought aloud.

"What's that?" Pauli said but Jung was too deep in thought to address that now.

"Are you saying we directly influence which field the energy resides in, even though our own physical bodies are being generated from the same quantum fields?"

"Of course! Our physical bodies are what allow us to interact with it at all!"

"I thought the energy in the fields would only be affected by something more fundamental. We have gotten so small that what influences things at this scale are things at a larger scale!" Jung said.

"Hang on," Pauli said, "the scale is not changing. Yes, things in reality and energy in the quantum fields are reciprocating cause and effect, but comparing the actions in our reality to the actions in the quantum fields is like asking if the chicken or the egg came first. You cannot have one without the other." Pauli explained. [FACT 90]

Jung thought on this for a moment and then clarified, "What you are saying is that the raw energy in the quantum fields cannot stay in its original form due to whatever laws of physics govern it."

"Right," said Pauli. "Let me add a layer of complexity to this whole thing. Remember superposition? What we see as superposition in human reality is that little cloud-like packet of energy where you can't exactly pinpoint the particle's properties or location in time and space."

"You can't pinpoint it because it is simultaneously being generated in all possible forms from the quantum field where it has just enough energy to stabilize in our existence." Jung felt like he was beginning to understand.

"And that's not all." Pauli said, sounding like he was selling something. "The fact that the very real particle in superposition acts the way it does is what allows it to be influenced by virtual particles so easily. When the particle is in superposition it is also a wave and waves build on each other with constructive interference. That is likely what's happening in the quantum fields as well." Pauli said.

"So perhaps, the subquantum energy is dancing on the cusp of reality until some other rogue excitation combines with it, then together, they can focus into a particle that can actually exist in our reality." Jung understood.

"Quite possibly, yes." Pauli said.

"So that's it then? You have just helped me answer one of my two core questions. I now understand what creates our reality down to the most basic building blocks." Jung took a deep breath before saying, "I continue my search for the ultimate depth of consciousness but now I have a new question. What is the force that spawns creation, that is, what is the energy beyond the quantum fields that is creating reality? Or perhaps the question is, what is the origin of the one single energy that is manipulated to create everything we know?"

"What is the consciousness beyond our everyday experiences and what is the energy of quantum creation? Two very good questions indeed." Pauli surmised.

Upon hearing his own questions aloud and with the example of a prism splitting light still in his mind, Jung was overtaken by an image of a white light shining like the shard of a crystal in total darkness. The pure light was radiating outward through the smallest field of one-dimensional points. It flowed through the sea of Planck sized points until it hit a quantum field like a prism splitting it into different colors.

The now brightly colored light shined through multiple quantum fields that acted as yet another prism; but this prism didn't just produce colors. Jung saw grand patterns of complex structures like a living kaleidoscope all originating from one pure light in the depths of reality. The white light was the pure energy of creation he was now hoping to understand. The colorful light represented the energy being hosted by each specific quantum field and the complex patterns represented the quantum fields working in tandem, creating protons and neutrons in our reality.

"The source energy, as you call it, is probably not being produced somewhere," Pauli theorized, "I think it just exists." He waived his hands as if clearing a cloud of smoke. "What originally created it, I can't say, but for the eternal present moment that is being generated by the quantum fields, I think the energy has always existed and instead of being created it is just changing."

"Sounds like God." Jung said quietly.

"Whatever made-up thing God is," Pauli retorted. "You know, the quantum fields themselves could be thought of as a function rather than a thing."

"A function activated by energy." Jung presumed. "Wait, does that mean that what you described as the quantum fields could just be a law of physics rather than a physical thing?"

"I equate it to photons not having mass but still being affected by gravity." Pauli stated. "A photon's velocity creates virtual mass. The quantum fields are not physical but their inherent ability to create particles requires them to be real."

Jung wasn't sure if he understood but he felt like he didn't need to; he had plenty to think about already.

FACTS

88.   Virtual particles cannot be detected for the reason described here. They do have the same amount of energy as a real particle (allowing them to be hosted by their quantum field) but they exist for less than one Planck time, therefore they do not exist and yet their influence has been proven by the Casimir Effect, and others.

By current standards the terms 'excitation' and 'virtual particle' are interchangeable. As physics evolves and we learn more about the unified fields of QFT, perhaps we will realize they should not be interchangeable.

89.   Mainstream science currently has a minimal understanding of a Planck Field with even less of an application for one. This scale is so small that even Quantum Field Theory starts to break down. The gravity that was

seemingly missing at the quantum level becomes very important at the Planck scale. This does not mean that QFT is wrong, just as QFT being correct does not mean Newtonian Physics is wrong. All we really know about the Planck "Field" is that we do not know very much. The depiction of the Planck Field in the narrative being a nice clean field hosting some type of pure energy is most certainly not correct. Reality at this scale is much more chaotic than this and science is currently working to realize the order within the chaos.

The author depicted this "field of fields" for two reasons. The first is to introduce the reader to the absolute smallest thing that could possibly exist: Planck length and time. The second reason is to give consciousness a medium, though this requires a departure from current scientific understanding that consciousness is only our waking experiences. If you are still not sure if consciousness requires any explanation beyond your experiences, then read on. When science finally acknowledges that consciousness (or some conscious-like thing) exists without the physical human body, it will require a medium to propagate it.

If there is one singular energy at the root of all creation it is probably more than one layer deeper than the Planck scale. Who among us will prove the next great realization of reality?

90.　　The point being made in the narrative is that the energy in the quantum field is literally the stuff in human-reality. They are inseparable as the old question poses; which came first the chicken or the egg; you can't have one without the other.

That is an example for the narrative, however, in reality, the chicken's egg came first. Evolutionarily speaking, there was one genetic mutation that led to the chicken's ancestor laying an egg which contained a chicken. So, the first chicken egg was laid and cared for by a creature that was not quite a chicken, thus the egg came first, that is simple evolution.

What is not so well understood about evolution is that external conditions can help force certain genetic mutations around the same place and time in many individuals. Meaning there was one, very first, chicken but its offspring might not have been the second. Instead, the second chicken could have received the same genetic mutation due to the same external factors at that time and place (without being genetically related).

We are, most of us, not special but beautifully average. There was a very-first person with green eyes but that does not mean all people with green eyes are decedents of that one person. [This is the author's opinion]

Green eyes are caused by two genes on chromosome 15: the OCA2 and HERC2. Mainstream science claims that one single person received a randomly mutated chromosome 15 and passed it on to everyone who now has green eyes. If one of us can receive a randomly mutated gene, then many of us can.

Furthermore, if there were external conditions that influenced a mutation it would make sense to see that trait emerge en masse from a single region and general time, which might look like a single person to researchers millennia later.

Chapter 29

Consciousness and Creation

"The deep critical thinker has become the misfit of the world, this is not coincidence. To maintain order and control you must isolate the intellectual, the sage, the philosopher, the savant before their ideas awaken people."

-Dr. Carl G. Jung

"So, the most fundamental part of our reality is a multi-field that exists everywhere like some nonexistent fabric of reality that focuses some other pure energy into all physical existence." Jung postulated.

"That is an oversimplified explanation, but it is not wrong. I will add that the way we study these fields is to describe different states they can be in." Clearly Pauli was unsatisfied with this oversimplified understanding of the quantum fields.

"Different states the particles can be in?" Jung asked.

"Actually, the particles are the different states the field can be in!"

"Oh, that is an interesting way to describe it." Jung said.

"It's an accurate way to describe it." Pauli said bluntly. "It was this understanding that allowed people to argue the fields might not be real at all, but they have been proven wrong. The fields are very real."

"I have another question," Jung said, "what affects excitations before they pop into reality? Or perhaps the question is, why do excitations build up sometimes while other times they just fizzle out?"

"We assume it is like energy in waves that combine or cancel out when they overlap, but no one really knows. To the best of our knowledge, energy cannot be created or destroyed even in the nonexistent realms of subquantum kinetics. All things are literally connected by the quantum fields. But to answer your question, perhaps God." Pauli shrugged with a smirk.

"Our individual conscious minds are in control of all the atoms that make up our bodies, it is almost like playing a small piece of God." Jung suggested.

Pauli nodded in agreeance. "That feels a bit blasphemous to be honest. Who am I to define God?"

"I think before this is all over," Jung said, "you will replace the idea of 'who am I' with the realization- 'I am, that I am.'"

This puzzled Pauli and after a few moments of silence Jung asked, "where does quantum entanglement fit into all this?"

"Well, entanglement is just when two or more quantum particles (inversely) share a property such as spin." Pauli said.

"Then entanglement only happens with particles that have already been created in our reality." Jung confirmed. "Could entanglement also happen to virtual particles? Or could excitations and virtual particles somehow *be* the mechanism that does the entangling?" Jung said with an excitement gaining in his voice.

"I could give different answers but each requires an interpretation that may be inaccurate," Pauli said, "so it is anyone's guess. As I mentioned at the beginning of this lecture we are left with only questions."

"Remember the theories I described as three branches on a tree?" Jung said as his mind raced two steps ahead of the conversation. "The theory of synchronicity could be explained by some connection of excitations in these quantum fields."

"I don't think I fully understand your tree. Now that we have been through all of general and quantum physics, what exactly are you getting at?" Pauli asked, genuinely interested.

"I feel that our consciousness connects us to something greater than ourselves, something that extends beyond the limits of our physical bodies." Jung said.

"I just don't see why you feel the need to find a definition for consciousness that extends beyond our everyday experiences." Pauli said stubbornly.

"Well for one, synchronicities, which only truly happen in the mind of the observer." Jung said.

"How do you figure that?" Pauli challenged. "In the story of the beetle, only one part was in the mind, there was a real beetle."

"Realizing there is a synchronicity *is* the synchronicity. Without the observer to realize the connection, the synchronous events are two totally unconnected things that would have otherwise gone unnoticed, like two ships passing in the night." Jung explained.

Pauli thought on this.

"Tell me," Jung continued, "does physics allow for meaning? What I mean is, could an observer's conscious thought physically effect any of that energy in the quantum field or does the math prove that could not happen?"

"This conscious thought you are referring to is when an individual assigns meaning to one of those synchronicities?" Pauli asked.

Jung nodded in affirmation.

"And that conscious thought is what you called an a-causal observation because the things seem connected but they were not the cause or caused by the other thing in the coincidence being observed."

"Exactly."

"Then to answer your question, although particle physics allows for an acausal form of observation, it actually has no use for the concept of 'meaning,'" Pauli said. [Real quote]

"Then it is neither proven nor disproven. It could be true but there is no mathematical need for a connection to exist." Jung thought aloud. "In other words, subquantum energy might get focused by random interactions; but is it possible that consciousness could actually focus, those rogue excitations, even if by accident or indirect means?"

"Perhaps, but why do we need consciousness to explain any of that?" Pauli argued. "It seems to me that the focusing of energy is more likely caused by normal interactions that follow whatever physical laws govern the quantum fields. There are no events that cannot be explained."

"The Pauli Effect cannot be explained." Jung said coolly.

Pauli gave a halfhearted chuckle. "How do I keep forgetting about my curse? The Pauli Effect is not a phenomenon in some quantum field, it is at the human level."

"You do not know that." Jung said authoritatively. "The coincidence happened in human reality but the synchronicity, noticed by the observer, where did that happen?"

"In their consciousness." Pauli reasoned.

"And where is that?" Jung challenged.

"It must be in the brain." Pauli argued.

"Yet we cannot point to a part of the brain and explain it. Consciousness may very well be a product of some unknown quantum or Planck force!"

Pauli disregarded this comment with a wave of his hand.

"And you just said no one knows what the laws of physics are at the Planck scale meaning no 'normal' has been established. So how could you possibly claim that only 'normal' interactions occur there?"

"I see your point." Pauli said, so calmly that it almost seemed out of character.

"What if consciousness *is* the law of physics that governs the energy in the Planck Field?" Jung suggested. "If any consciousness was able to focus energy in the quantum fields then all those mystics would be right. Thoughts could literally create reality. And you just told me that no math or science disproves this possibility! Some aspect of your own mind might focus those subquantum energies to cause things around us to manifest in certain ways. In your case malfunction!"

Pauli was stunned, his logical nature was to find a scientific solution but with the thought of the most recent Pauli Effect in his mind, he felt in his heart the scientific method was of no use here. "You might just be on to something Doctor Jung." He said though he still didn't agree with the wild ideas Jung was proposing.

He sat and pondered for a moment on how he could get Jung to admit his argument was folly. Then he got an idea. "I have now arrived at one of your two fundamental questions so now I have a question for you. It is the reciprocal of the question you just asked me. Synchronicities are simple coincidences that are odd but do not break any laws of nature. Is there any actual evidence to suggest consciousness exists outside the human body where a law of physics was broken?" Pauli sat back in his chair thinking he was about to get an easy victory.

"Perhaps there is." Jung said mystically as Pauli shot upright in his chair. Now Jung was the one smirking at Pauli's ignorant curiosity.

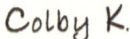

Chapter 30

My Odd Cousin Helly

"Imagination is more important than knowledge. For knowledge is limited, whereas imagination embraces the entire world, stimulating progress, giving birth to evolution."

-Albert Einstein

"My mother's side of the family were all," Jung paused, "a bit odd."

"Odd how?" Pauli blurted out as he was becoming desperate to hear what evidence Jung might have to support the claim of consciousness outside the human body that also broke some laws of physics.

"My grandmother was known as a spirit-seer with clairvoyant abilities." Jung explained. It seems that she passed these abilities on to others in the family, including my mother. When I was very young, I saw things and heard noises around my mother. One night, I must have been six years old, she was lying in bed humming in a trance and as I entered the hall a headless specter walked out of her room. It stood facing me and a light appeared where its head should have been. The light rose above the body and disappeared."

"You have always struck me as a man of science." Pauli challenged. "These ghost stories sound like child's play. Do you really believe you saw a ghost?"

"I saw something, and I agree, let's look at this as scientifically as possible. What I saw was either my imagination, or it was in some way real. Only one of those two possibilities can be true. If it was imagined, then science can already explain everything that occurred. Light may have tricked my young eyes and my immature understanding of the world would have influenced my perception of the event."

"Or it was real." Pauli said.

"I'm not arguing that ghosts are real, I'm suspending my logical disbelief so that I may ask, if that memory is accurate then how can I explain it scientifically? If I saw a bona fide apparition, then I may have witnessed consciousness in extra-corpus form."

"Consciousness outside the physical body. That would change our entire understanding...of everything." Pauli said cautiously. "To change an accepted theory, all science needs is one single fact that contradicts the rule. But one instance like this would not constitute a fact at all."

"Right." Jung agreed, "my experience is one single data point. Far from a fact."

"We cannot even consider it a scientific observation. And how would you go about measuring it?" Pauli challenged.

"Exactly the conundrum I am facing." Jung admitted. "All we can do is describe the conditions that were present leading up to, during, and after the event while counting the whole thing as one instance."

"Ahh, I see." Pauli said. "Then over time and after interviewing many people who had similar experiences you will eventually be able to say something like, in 1,000 cases of apparent extra-corpus conscious apparitions, 342 of them occurred during a certain condition."

Jung knew what Pauli was about to suggest and it felt like a weight on his shoulders, but he allowed Pauli to finish anyway.

"Then, one day, we could create those conditions and replicate the findings in a laboratory experiment."

"Unfortunately, I have already done that and it didn't work." Jung admitted.

"What!?" Pauli exclaimed. "You have already attempted a scientific experiment to conjure some extra-corpus consciousness?"

"Allow me to explain. It started when I was around 20 years old. It is my cousin, Helly, whom I am now speaking of."

"She also inherited these strange qualities from your grandmother, I assume." Pauli said.

"Indeed, she may have been the oddest of all. One summer we began to hold séances where family members, mostly Helly, would communicate with spirits. We would come together every so often until I started to pull away. I felt that if I continued attending then I would not fit into society, and the normal life I dreamt of for myself would slip away and I would be labeled a heretic."

"Not good for someone in the medical field to lose credibility." Pauli agreed.

"For that and another reason I stopped attending the séances and my absence caused them to cease altogether." Jung explained.

"Wait, what happened during these exchanges? Did you see spirits?"

"Not really." Jung admitted, "my cousin would change her voice and speak as if she was different people, presumably spirits."

"That's it?" Pauli couldn't hold back his disappointment. "That's your big evidence?"

"Not even close, I've barely begun. She produced knocking noises in furniture and walls up to 20 feet away from the closest person. On one occasion I tested it, as I had a healthy skepticism for the whole thing.

"I sent Helly to her house about 4 kilometers away with the intention of producing these noises where I was. About the time it would take for her to get home, knocking noises were produced all over my house. I ran inside and outside, trying to find someone playing a joke on me but I was totally alone." [FACT 91]

"Interesting." Pauli tried to give his friend credit and find some tangible evidence in the story but came up with nothing testable. "Did anything else happen?"

"When she communed with these spirits, she did seem to possess knowledge that was beyond her. Sometimes she said things that had been personal secrets of people that we later found out to be true. It is possible Helly independently realized these people's secrets. There was never any definitive proof besides odd thumps in the night."

Jung sighed before continuing, "Once, though, I felt like I caught her cheating in one of the séances. I challenged something she said in her trance and as that séance was ending, she announced, 'It was Carl. He has to leave. He doesn't deserve the flowers.'

"The séances stopped and I resumed a normal life until the following year when I gave her a birthday gift. It was a book about seers. She unwrapped the gift and read the title aloud. She immediately burst into a spontaneous psychic trance. She said, 'Carl wants to explore the soul and the hereafter.' At the time I was studying to become a surgeon, I had no interest in those things. I disregarded it as her seeking my attention but I always had this nagging feeling she was right." [FACT 92]

"This still all fits within our known laws of physics." Pauli said.

"It was another two years after that. It was my first day home on a visit from medical school when it happened. The kitchen table where we did the séances was made of hard oak and had been in the family for 70 years. Helly and my mother were in the room and as I sat down something happened that will remind you of your own Pauli Effect. There was a massive bang, like an explosion, and the round table split from the center to the outer rim, directly towards my cousin."

"So, you are akin with the old Pauli Effect, I see!" Pauli said with a chuckle.

"Perhaps I am," Jung continued. "Helly suggested it was a sign from the spirit world to begin our séances again but I was skeptical. I was there for two weeks and no other strange things happened until my last day. It made me realize my skepticism had been misplaced."

"Yes, what happened?" Pauli said impatiently.

"It was late and I was eating an evening snack, country bread. After cutting the bread, I set the knife down on the cloth that wrapped the bread. Another loud bang occurred, so loud it woke my neighbor. The knife shattered; the blade broke into four pieces." [FACT 93]

"The knife you had just used?"

"Yes."

"You must have done something, left it on a hot stove, perhaps?"

"Please tell me about a time when you or anyone you know had a bread knife explode in your kitchen. Even if it had been left on a hot stove that would not cause it to explode, this is unheard of!" Jung blurted out. "It is beyond normal explanation!"

"Yes, very well." Pauli relented. "Was your cousin present when this happened?"

"No. I interpreted this to be a sign from the spirit world directed at me, just as Helly suggested. This caused me to resume the séances which produced the same results and that is when I gained the courage to invite colleagues from the university for a proper experiment. With them present, nothing was conjured and Helly could not enter her trance as usual."

"Fascinating. Bumps in the night to exploding kitchen items. Well, I am just glad to hear that I am not the only one who suffers from the Pauli Effect! It seems to me that one of two things is true. Either all of the séances and psychokinetic things are real or they were all faked. If faked, then we have nothing else to discuss."

"Right." Jung agreed.

"However, if just one of those things was true then we have a scientific observation that contradicts our current theories and therefore our current theories are wrong."

"Or at least incomplete," offered Jung. "It could be an opportunity to learn something new."

"Overturning a scientific theory is no small matter." Pauli said.

"As far as science is concerned spirits don't exist at all, so any evidence of a spirit goes against the scientific community's understanding of reality." Jung reasoned aloud.

"Let's dissect both possibilities." Pauli led the discussion. "Many odd and unexplainable things happened but when it came to reproducing the results with scientists present everything failed. Was your cousin making it all up somehow?"

"I don't think so." Jung spoke honestly. "It is possible though, and if that's what Helly did, then her motivation was simply to spend more time with me. But similar things happened to, or because of, my mother and that ended with her involuntary hospitalization. She had no motivation I can see and would not have carried on a charade to the point of being... imprisoned."

"If we assume these events happened just as they seemed, then how do we explain it?" Pauli asked. "And why didn't those things happen during the experiment?"

"Perhaps my mother and cousin's psyches could produce certain conditions that are required to achieve those results, but those prerequisite conditions could not be met because the doctors' psyches negated them." Jung's comment seemed to distract Pauli.

"Destructive interference," he mumbled curiously.

"You taught me that results of quantum experiments change based on how the observer measures the results. What if the doctors unknowingly changed the conditions?"

"I must remind you that it is not the conscious observer who influences the properties a particle obtains."

"Yes, and I am saying that there may be an inherent property that changed the outcome in this trial just like the act of detecting a particle changes it. Did you know that up until the 1920s, some pilots reported their instruments acted improperly when they were in a high emotional state? This might demonstrate exactly the same thing I am saying. Someone later did an experiment which showed the pilots' emotions did not have any effect on the instruments and from then on there were no more reports of high emotional states influencing the instruments."

"So what?"

"Don't you see?" Jung exclaimed. "This all adds up to exactly what those mystics talk about with the law of attraction!"

"How so?" Pauli said defensively.

"We get what we expect. When the pilots expected their emotions to manipulate the instruments they did. When my family expected to produce supernatural events, we did. The scientists in both experiments expected to find nothing, it was their outright intention! After they published their findings no one expected high emotional states to manipulate mechanical instruments. It is almost like a barrier was created, blocking what had previously been some parapsychological connection."

Pauli scoffed. "Expectations mean nothing! Experiments happen all the time where the results are not what was expected."

"Yes, I agree there is more to it than simply expecting a certain result but how will we find out what that is unless we acknowledge something currently unexplainable is going on?" Jung challenged. "In this type of experiment, we are looking for results produced or influenced by the subject's consciousness. Therefore, it is not unreasonable to say the scientists' consciousness may have affected the results. When the scientists weren't there the results were deeply personal and the people were emotionally connected to the odd results."

Pauli thought for a moment before curiously saying, "Go on."

"Think about the pilots," Jung explained, "their lives depend on their instruments, so their emotional state is directly connected to them."

"Entangled," Pauli mumbled.

"My family's odd experiences were deeply personal to us. In each instance there was a deep emotional connection to the observed phenomenon."

"But when the scientists came to run an experiment, they were emotionally connected to their scientific beliefs which require ghosts to not exist," Pauli suggested. "And their emotional state interfered with your family's which broke or prevented the connection. Could emotional connections influence reality?" Pauli lightly asked aloud in a curious tone. "Ah ha! I see it, I see how you arrived at that core question of yours. What is consciousness?"

"Now you've got it." Jung said.

"It is like there is a missing piece to this explanation and it seems like it could be something like consciousness though I would wager it's something entirely different, though somehow connected or similar." Pauli admitted.

"I still feel skeptical of all this," he continued, "but I will admit I'm intrigued. I appreciate how you reduced your experience to a single data point and I see now that you have many of these data points, not a large enough sample size to prove anything, but it is enough to get my attention.

"In your first experiences," continued Pauli, "you were quite young and could have been mistaken but, in the others, you were mature enough to draw accurate accounts. The only question there is, did something outside of our known reality crack the table and knife or did your cousin dupe you."

"My cousin is smart, but I don't believe she could set up something that elaborate. As crazy as it sounds, I believe the most likely explanation is some parapsychological phenomenon. Just think about the sheer strength required to split that table, and with us all sitting right there; we didn't see any tools or anything."

"The other option is that the table and knife have other explanations within the laws of physics, but we are missing some information."

"No," Jung responded, "I cannot fathom any explanation within the known laws of physics. These events happened on my first and last day in town which also adds to the coincidence."

"Then perhaps it was not your cousin but you who conjured up the necessary conditions for your table and knife to break." Pauli suggested before his tone turned bleaker. "I appreciate your conviction, though as we have said before, I do not see how any bit of this can be proven."

"Yet." Jung said with a twinkle in his eye.

Pauli paused and his friendly smile soured into a grimace. "I feel you are forgetting that we proved the conscious observer does not directly influence the outcome of the experiment. I, myself, led the experiment. We shot single photons through the double slits and recorded which path each photon traveled. We then destroyed the recordings of the test group without ever allowing any conscious person to view them, but the results were the same."

"Hubris, Herr Pauli!" Jung said confidently.

"I beg your pardon?" Pauli replied with surprise as his temper moved to the brink of a flair up.

"That experiment did not prove anything about consciousness." Jung explained.

"And how do you figure that?"

"We do not know the limits of consciousness, so how could you possibly claim you proved it is not affecting your results?" Jung challenged.

Pauli realized his experiment was designed on the assumption that consciousness is limited to our everyday experiences.

"You have shown me that our minds are more than our waking consciousness, but you haven't proven that consciousness exists outside the human body."

"Think of the missing piece you just described," Jung said. "A force or energy or perhaps even a law of physics; you said it could be consciousness or something else entirely, though somehow related."

"As scientists we can only test what we have access to. How would you propose we test for this extra-consciousness then?" Pauli snapped.

"We can't Herr Pauli, that's why it is so frustrating!" Jung said. "No one knows the limits of consciousness, so no one has proven it is or is not affecting quantum particles." [FACT 94]

"Hmpf." Pauli grunted.

"I hold that some form of consciousness exists outside of the physical body. Perhaps even, a consciousness different than our individual self. Just because your own physical body did not witness or observe the results means nothing. Did you feel a deep emotional connection to the experiment when you were conducting it?" Jung challenged.

"Well, isn't that convenient for you." Pauli said calmly in his raspy voice, before his voice began growing louder. "You think the Holy Spirit observed the experiment and triggered waveform collapse. What other ghost stories do you have for me, Herr Jung?" He yelled.

"Calm down, my friend," Jung said, waving a hand at Pauli. "There is a question someone once asked me, and I think I've finally answered it. If our own consciousness somehow influences or creates reality, as the law of attraction wizards believe, then why do bad things happen to good people?"

"And?"

"And I now understand there is a collective consciousness that we somehow interact with. We can influence it." Jung explained. "The missing piece you just described; I believe that is this."

"What are you saying, you and I are God?"

"That is an interesting way of putting it." Jung pondered for a moment, "I suppose that is exactly what I think. I nor you are God but collectively we are all God, or at least we can tap into God's essence via our individual consciousness." Jung said. "Bad things happen to good people, not because of our individual influence on this collective mind but because of everyone's influence."

"And who belongs to this mind? Is it limited only to the humans on Earth or does this unconscious hive-mind include dogs and cats too?" Pauli went on a tirade. "What if there is some alien life deep in space, are they part of this singularity too?" Pauli was frustrated with Jung's imagination and how none of what he was saying could be tested.

"Truly, it has no limit." Jung said softly with conviction. "The animals, aliens, grass, the trees, even the wind and the rocks, gravity, and the likes, literally everything either is, or contributes to, this collective consciousness. All people and things from the past and all people and things from the future would have an equal influence on this great collective."

Pauli remained silent as Jung continued.

"When something happens, it has a ripple effect throughout all of space and time, every action, every existence contributes to our objective reality. When one change is made, that moment will affect things generations later, even if the people have no idea of that event."

"The butterfly effect." Pauli mumbled.

"Take for example a flat tire. If a man was driving and one of his tires blew out, causing a wreck that killed him, and his surviving family could no longer afford to live there, then three generations later they don't feel affected by that flat tire at all."

"In that case everything that ever happens to the person generations later only happened as it did because of a flat tire in some other place long before." Pauli said. "Interesting."

"I think we have these ideas of good events and bad events; but in reality, and perhaps quantum science agrees, there is no good or bad, only events."

Pauli chuckled, "I think all of science agrees there is no good or evil, only stimuli and responses. You have experienced many strange things, but I feel like there is more that convinced you to study parapsychology."

"I think that was you Herr Pauli!"

"Me? How?"

"The Pauli Effect fits the same description of all the other odd things I've come across in my life. Before my cousin Helly, I saw my mother conjure up spirits; I saw something that night, I felt it too. Before the Great War, I had a series of premonitions of what was about to happen. Where did that knowledge come from?"

"It became personal to you."

"Exactly, and finally a highly logical physicist happened into my office and told me about how his presence breaks things. It all fits a similar pattern. I know things happen that cannot be explained by modern science and for some reason I am charged to find the explanation; it might be my life's work just as Helly predicted."

Jung leaned back into his chair as silence fell upon the office.

The silence burned like a fuse and just as it became uncomfortably long Pauli finally said, "It appears you have won today's boxing match. I cannot deny my own deeply emotional connection to every single instance of the Pauli Effect. If there is more to consciousness than our own waking experience, then perhaps I did not prove the observer has no influence on the results of an experiment."

"We find ourselves again asking; what is consciousness?" Jung gently said.

"Dammit all. What are the limits of consciousness?" Pauli said as he leaned back in his chair.

"Perhaps the psychologist will have to become a physicist in order to find out." Jung said.

"Perhaps the physicist will become a psychologist." Pauli said with as friendly a smile as he could muster. "Perhaps I should enroll myself as a student at the Jung Institute. I've heard rumors it is about to open here in Zurich."

"Ah yes!" Jung said, jumping to his feet before rummaging through his desk. "The rumors are correct Dr. Pauli. It will finally come to fruition after nearly a decade since I first conceptualized it. And thank you for reminding me! The opening will be celebrated with a banquet this Saturday evening. I would like to personally extend this invitation to you." Jung said as he handed Pauli a small envelope.

"I will be there." Pauli said humbly. "I appreciate your unique perspective, on the nature of the universe and my place in it."

"I am glad you feel the counseling is helping." Jung replied.

"Until our next meeting, Doctor Jung."

"Good day, Doctor Pauli."

FACTS

91.   Carl Jung did have an odd cousin named Helly. He did attend seances with her and other family members as described in the narrative. She would often speak as if she was speaking for the spirits of people who had passed. They regularly heard strange knocking noises that had no explanation. On one occasion Carl did set up the experiment where he sent her away challenging her to produce the knocking phenomenon at his house. It happened just as described. [Ref 3]

92.   Jung recounted these facts in his book "Memories, Dreams, Reflections" and the depiction in the narrative is accurate to his memory. [Ref 3]

93.   The table and knife did crack and break as described here. Jung even provides photos of the knife in his book, "Memories, Dreams, Reflections." [Ref 3]

94.   Pauli never led an experiment to test consciousness's effect on waveform collapse. However, others have. They claim to have tested for consciousness affecting quantum particles, but their definition of consciousness was limited to our conscious experiences, not including our unconscious mind or any other, currently undefined, aspects of our consciousness. All that has been truly proven is that human's everyday conscious experience does not affect quantum states. [Ref 74, 92]

There is currently no scientific belief that consciousness has any ability to entangle particles. However, if any of the parapsychological events described in this book did happen as Jung believed, then some form of entanglement could be a sensible explanation.

My Odd Cousin Helly

Chapter 31

Symbolic Language

"Your visions will become clear only when you look into
your own heart. Who looks outside dreams,
who looks inside awakes."

-Dr. Carl G. Jung

April 24th, 1932

The sun slowly sank into the evening sky as Dr. Jung entered a lovely Victorian style building, complete with a turret on one corner. The steep roof poked the sky, four stories above the street below. There were as many windows as would fit on the building, each with hand painted shutters. After a light renovation the building became the C.G. Jung Institute of Zurich at 27 Gemeinde St. It was a modestly sized building for an institute but had plenty of room for the few post-graduate classes it offered. [FACT 95]

Jung arrived early to the institute's commencement banquet and found a few of his staff already preparing for the festivities. His administrative assistant, who handled all matters from admissions and transcripts to printing diplomas, placed a blue and white vase on a small table in a corner of the room. She set it down a little harder than she meant to, and it wobbled for a moment as if there was something underneath it.

"What lovely blue irises you've chosen for this evening." Jung said to her.

"I'm afraid the flowers won't make it past this evening. I accidently let them dry out in the hot afternoon sun. I've just filled the vase with ice cold water though." The assistant mentioned casually as she brushed a tiny rock out from under the vase, allowing it to sit flat on the table.

Perhaps it was caused by the ice-cold water shocking the hot vase or perhaps it was the little pebble the assistant didn't realize she set the vase on. Either way, a tiny, unperceivable crack formed in the vase, which now sat in a lonely corner of the room.

Jung placed the speech he prepared on the podium in front of the head table as the first two guests walked in.

"Willkommen Professors." Jung said as he recognized the dignitaries from the University of Zurich.

Guests trickled in until the room was full of chatter. Jung scanned the room to take attendance and noted he did not see his friend Pauli. While scanning the room, out of the corner of his eye, Jung noticed a stocky man barrel through the door but before he could look to see who walked in, he was distracted by something on the other side of the room.

A loud cracking sound followed by the shattering of some window or possibly a ceramic dinner plate quickly silenced the room as everyone looked around to see what the commotion was. Jung looked to the corner of the room where the blue and white vase had been sitting and saw that somehow it broke. No one was near it before it spontaneously shattered. Without looking to the door, Jung knew beyond any doubt that his friend had just arrived in his most unique way.

Wolfgang Pauli stood perfectly still having just crossed the threshold into the banquet hall. Jung stared into his face and smiled as he recognized the Pauli Effect had just struck again. [FACT 96] Pauli walked to the corner to inspect the vase.

A voice from across the room exclaimed, "Pauli just walked in, no other explanation needed!" A few people from around the room laughed, some because they knew of the Pauli Effect, and were intrigued they got to experience it firsthand, while others laughed only because they were aware of his callus reputation.

There was a large portion of the blue and white vase still on the table but some had fallen to the floor and shattered. Pauli stood pensively staring at the water slowly pouring out the broken shards of porcelain. He wasn't sure why but he was entranced by the water as it dripped out of the vase and onto the floor.

"That was an entrance perfect for the occasion Professor Pauli! I'm glad you could make it, my friend." Jung said, shaking Pauli's hand.

The banquet began and the sound of silverware clanging against dishes was heard over the conversational chatter that filled the room. As plates were cleared away, Jung took to the podium to provide his commencement speech.

"Dignitaries, colleagues and friends, it is with great pleasure that I am sharing this moment with you. I have found myself as part of this century's zeitgeist. From Vienna to Copenhagen to right here in Zurich, and many places in between, it is us who are delivering the world into the quantum age.

"I am going to speak to you this evening about some of my more complicated theories which will attempt to describe the foundation of the human psyche and may, one day, help describe the foundation of physics.

"The human psyche has its own language. More similar to the hieroglyphs of ancient Egypt or Central America, this language of the psyche is symbols. Dreams permeate messages to us from our unconscious mind via these symbols. Likewise, when we read a good story or watch a play we relate to characters because they are archetypes, symbols written in the language of our unconscious minds.

"There is another type of symbol I would like to discuss this evening. These are the living symbols that present themselves to us in our everyday lives. If, for example, you become inexplicably fascinated by a specific object or scene playing out before you, even if you are not actually interacting with it, it could very well be that you have noticed this thing because it is a symbol representing a message from your subconscious mind in the same way a dream would.

"Each of us have an internal world and an external world. The external world is logical and social while the internal world is emotional and instinctual. It is this internal emotion that can connect us to symbols we see in the real world."

Pauli was immediately distracted by this idea and shot a glance back to the broken vase as Jung continued his speech.

This time, Pauli was not concerned with the Pauli Effect. Instead, his thoughts lingered on the fact that he had momentarily become entranced by the water that was dripping out of the broken vase exactly how Jung just described. For the first time since he was a child, Pauli dropped his logical defenses and allowed himself to feel the connection he had with that broken vase.

He dwelled on it. He pictured the water pouring from the vase and felt something odd. A warm sensation grew in his chest as if the feeling in his heart was opening his mind to untapped knowledge. It was like the universe was trying to show him something, as if there was a message just for him. This is one of those symbols Jung is speaking of, Pauli thought to himself. If that is true, then what does the vase represent in my life, and what about the water?

How would physics explain this inherent property that he is describing? This would mean that anything could be a synchronicity if you wanted it to be. The only explanation physics could give for this would require the whole universe to be like some intertwined matrix of energy with entanglements everywhere. If that is true, Pauli thought, then the whole universe operates with some sort of background physics and is ultimately an expression of a singularity, like the inverse of the densest black hole imaginable, whatever that is. The point where our consciousness meets the collective would be an event horizon, which may inherently prevent us from studying it... ever.

Pauli had a feeling in his chest. It felt vulnerable but something about it was comforting. He explored the odd feeling and somehow, he felt connected to everything around him. The feeling seemed nostalgic, like an innocent memory from childhood mixed with the strange sense of déjà vu. As he grew up his brain learned to differentiate himself from the world around him, but in this moment, all those barriers were gone and he was as free as a child, connected to everything around him.

Pauli didn't hear most of his friend's speech, until he tuned back in at the very end.

"I officially declare the C. G. Jung Institute open!" Jung announced to enthusiastic applause as the entire room celebrated this new beginning.

FACTS

95.   The inauguration of the C.G. Jung Institute was a real event that took place on April 24th, 1948. Jung planned on creating the institute almost a decade prior but was forced to wait due to the Second World War. [Ref 75, 76]

96.   A vase did break at this event just as Wolfgang Pauli walked in. The vase breaking in real life had no explanation. [Ref 76]

In the narrative, there was a small crack and temperature shock which was the cause and it is only the timing that was synchronous. If we investigate the parts of a synchronicity and find each part has a logical explanation, does that negate the synchronicity? If we know why the vase cracked and it had nothing to do with Pauli then the timing is a pure coincidence, right?

Synchronicity is when we notice a coincidence and can find meaning in it that applies to our own life. Even if we find the logical causes of the parts of the synchronicity, it was still an event that we were tuned-in to experience in that moment. What synchronicities have been happening in your life that you logic into coincidence?

Chapter 32

The Coins are Conscious

"There is a mystic fool in me that proved to be stronger
than all my science."

-Dr. Carl G. Jung

Wolfgang Pauli arrived promptly at his scheduled time for another
week of counseling.

"Good afternoon, Dr. Pauli." Jung greeted.

"Hello to you." He replied.

"I figure we can start by discussing your dream, has it changed at all?"
Jung tried to begin the session.

"When you were talking about the language of the human psyche,
what did you mean?" Pauli blurted out, totally disregarding Jung's question.

"Oh, you mean from my speech the other day." Jung realized.

"Yes, let's hear it," retorted Pauli.

"The human psyche has a universal language." Jung explained. "It is
not words but symbols. Basically, all humans in the whole world speak that
same language with different dialects broken down by culture."

"There was more to it." Pauli said, unsatisfied with that answer.
"There was something, almost mystical, about your explanation."

"What if the conscious energy that allows us to understand a symbol in a dream is the same energy as the symbol in real life?" Jung proposed.

"What does that mean?"

"Everything you've ever experienced happens inside your mind." Jung stated.

"How do you figure?" Pauli asked.

"There is an external objective world we live in and interact with but we are each creating an individual simulation of that world inside our minds." Jung explained. "Look at the painting over there." He pointed to a large square painting that was just a circle with a bunch of lines forming some odd pattern. "You see the red color?"

"Yes."

"Well, there is an objective world out there and the red color exists as it truly is; but…" Jung said dramatically, "how do you know you are perceiving the exact same hue of red that I am?" Jung asked.

Pauli reminded himself how the eyes work by detecting photons and sending information to the brain. "The photons obtain the red's frequency when they reflect off the painting. The frequency of the photons hitting my eyes are the same frequency as the photons hitting your eyes, so we are both perceiving the same shade of red." Pauli reasoned.

"What about someone who is colorblind?" Jung challenged, "or for that matter, someone who is color deficient or perhaps just has old eyes?"

"Well, that would not change objective reality." Pauli argued.

"That's right, the thing in reality would not be different, but some people would have a different conscious experience given the same stimuli." Jung explained.

"Okay," Pauli relented, "what is your point?"

"You can only see, smell, and feel things inside your mind." Jung explained. "Your physical body receives inputs from the external world but you don't experience the external world directly. Your senses translate the stimuli outside and send electric signals to your brain."

"Yes, I understand this concept," Pauli said. "Then the brain reconstructs reality inside your mind." Pauli thought on this for a moment before saying, "but I see your point. My conscious experience is not capable of directly interacting with objective reality. It can only use my physical body to experience the external world." He remembered his original question about the mind's language and said, "what does any of this have to do with that bit about the mind using symbols as language?"

"There is an internal world and an external world, but to our brains they are the same thing." Jung explained. "Did you know there is no discernable difference in the brain between a memory and imagination?"

"Memory and imagination are the same thing?" Pauli questioned. "How could that be?"

"They are not the same thing but they are both very similar processes that occur in the mind. If you were to watch the electrical signals in a brain as the neurons are firing, you would see the same parts of the brain light up when someone is remembering a real event or just imagining something totally fictitious."

"That is strange," Pauli said, though he could not tell what point Jung was making. "I always thought a memory was stored in the brain like data on a computer." [FACT 97] "I still don't see how hieroglyphs fit into all this?"

"I'm getting to that," Jung said. "When someone is dreaming it can also look the same as when someone is experiencing life; parts of the brain are shut off, but they are similar looking processes." He explained. "Therefore, if you dream of a dragonfly and it carries a symbolic meaning, it could be the same if you notice a dragonfly in the real world. It too, may carry the same symbolic meaning."

Pauli gave Jung a questioning look, finding it hard to believe the objective world could provide symbols to people the same way a dream does.

"If that sound too out there," Jung said confidently, "allow me to demystify it for you. Think back to when you and I were walking through that park to the café. Do you remember seeing the owl swoop down to the bush to our right?"

"No." Pauli said.

"But you were looking directly at it. Why did I notice it, but you did not?

"Perhaps I was preoccupied describing my dream," Pauli reasoned, "or maybe at that exact moment I was looking away."

"I was on your right. I only noticed the bird because your eyes darted up over my shoulder. You may not have 'seen' the thing, but photons bounced off it and hit your eyes."

Pauli searched his memory and remembered Jung on his right but still had no recollection of an owl. "Okay, there might have been a bird I didn't notice and you did, so what?"

"Think about the process of sight. Our eyes collect raw data and our brains piece together the information which then creates the visions we see."

"Photons hit the eyes, the optic nerve sends electric signals to the brain, etcetera." Pauli said hurrying the conversation to get to the point.

"Heuristics Herr Pauli, you cannot forget the heuristics!"

"Okay, the understandings our mind implies on the information we perceive. What's that got to do with this?" He said impatiently.

"Everything!" Jung said, frustrating Pauli. "In any given moment we are seeing trillions of photons. Our eyes observe everything in our field of view but our conscious mind does not *notice* everything. When the electrical signals from our eyes are received in our brains, they don't just instantly become images." Jung explained. "The brain has a process to create what we experience as sight. Do you remember the light-comes-from-above heuristic?"

Pauli did remember, "right," he said, "a process the brain uses to understand the external world quicker and more efficiently."

"There is another heuristic."

"Okay," said Pauli.

"The unconscious mind focuses our attention on what it thinks our conscious mind needs to see at that moment." Jung explained. "That is what we perceive as our conscious experience!"

"That doesn't make sense, when I see, I see everything." Pauli said stubbornly.

"Then prove it." Jung challenged.

"Okay, how?"

"Remember that window you broke a couple weeks back?"

"Yes." Pauli said, shooting a glance to the window which still had a temporary board covering it.

"We both looked at it after it broke, didn't we?"

"Yes."

"How many panes of glass were in that window?"

"This is a game," Pauli said. "What does it matter?"

"You don't know, even though you definitely looked at it. Your mind never showed you that information even though your eyes saw it, just like the owl in the park." Jung claimed.

Pauli glanced to the smaller window behind Jung's desk. It was a totally different style than the large broken window. He closed his eyes and searched his memory for the window in question. In his mind's eye he saw four large panes, though the bottom left one was shattered. "Four." He said confidently.

"Wrong." Jung said calmly. "There were six."

"Okay, so what?" Snapped Pauli. "That information wasn't important enough for me to remember it!"

"The point is," Jung explained, "sometimes your unconscious mind chooses for you to consciously notice certain things."

"Why would it do that?"

"Sometimes for survival, sometimes for the same reason it would show you a symbol in your dream!" Jung explained.

"What are you saying? Somehow our minds are constantly sending us subliminal messages?" Pauli asked.

"Not subliminal, symbliminal." Jung said mystically. [FACT 98] "It's when stimuli in human reality are consciously observed by an individual and that thing carries a pertinent symbolic message."

"Hmm," Pauli thought, "okay, this is getting interesting." He relented.

"Have you ever noticed something and somewhere deep in your gut you felt like you were supposed to notice it, as if it had a message for you?"

Pauli was violently transported back to the banquet where he was staring at the water slowly dripping from the broken vase. That is exactly why I asked this question, he thought.

Jung continued, "that was your unconscious mind calling attention to a symbol so you could figure out what the message is."

"One hundred thirty-seven," Pauli mumbled as the idea of other symbols he regularly noticed in objective reality popped into his mind. "But you aren't saying that somehow the unconscious mind is literally creating those things in human reality?" Pauli clarified.

"What I am claiming is that when you dream of a symbol or you see that same symbol in reality, *and really notice it*," Jung emphasized, "both instances could carry the same symbolic meaning."

"And since the dream and the real thing are both being pictured inside your brain there really is no physical difference." Pauli realized.

"Does our unconscious mind somehow manifest those symbols into reality; you ask." Jung continued. "The answer to that question depends on quantum physics. As I understand human reality, probably not. However, as you have explained the quantum world where the most minute building blocks of reality can be connected across vast distances, I have started to wonder what role consciousness plays in all that." Jung explained.

"Yes, but I have told you, there is no need for consciousness to be any part of it." Pauli reiterated.

"You have said that, and yet here we are, two conscious people, consciously talking about consciousness with no explanation whatsoever as to where consciousness comes from." Jung said cooly.

"Ha." Pauli scoffed.

"There are parts of the brain we can point to and say this pattern looks like consciousness but there is nothing inherently special about it. There is no physical particle that is, by itself, consciousness bouncing around our minds. Are you now claiming that consciousness doesn't exist at all because the math doesn't have a need for it, despite your own conscious experiences?"

Pauli took a deep breath. He was introspective for a moment then said, "you know, math isn't always right."

"What?" Jung was shocked.

"There are mathematical fallacies where a math problem works out correctly but when applied in physics, or the real world, that same equation can be proven as objectively wrong." Pauli explained.

"How is that?"

"Mathematics are just a way to describe reality." Pauli began.

"Right, you described it as the language of the universe."

"Your dreams aren't the only thing symbolic around here. I symbolize aspects of reality as letters on a chalk board. Sometimes an equation can work out but different people interpret the symbols differently. A major one in quantum physics is where we have the Copenhagen interpretation, the many worlds theory, or the de Broglie-Bohm interpretation. Different sets of mathematic equations all related to the same observed phenomenon, interpreted vastly differently, each with its own unique supporting equations. Each set appear to work out correctly."

"But only one can be true." Jung finished the idea.

"Therefore," Pauli continued, "it is simultaneously a mathematical truth and an objective fallacy."

"Then mathematics cannot be the standard I thought it was. I've been thinking it was the ultimate way to prove something but you're saying it can be wrong." Jung realized.

"Whatever truth you are searching for, you probably just need a new perspective or a deeper understanding of something else first." Pauli echoed the sentiment he had instilled in Jung.

"Like consciousness," Jung said, "I think I need a deeper understanding of quantum physics for example."

"Ha!" Pauli laughed, "then perhaps we should continue our discussion on the quantum fields, but first, how did you come to understand this internal mind and external reality?"

"Wilhelm," Jung said, leaving Pauli to wonder whether he was referring to the king or someone else. "He introduced me to an ancient Chinese knowledge." Jung said mysteriously.

"Sounds exotic."

"It's called the Book of Changes. I think most scientists today would consider it akin to divination or reading tea leaves, but I have found a scientific use for it." Jung explained.

"How is that?"

"Someone can flip a set of coins then take the results and find insights into the question they posed."

"Like a fortune-teller." Pauli said.

"Kind of like reading a fortune." Jung relented.

"Honestly I'm a little disappointed." Pauli said realizing there was no scientific knowledge.

"Hang on," Jung protested. "Why has this practice been around for thousands of years? Kings have lived and ruled based on the outcomes of the coins."

"And in the Middle Ages kings had wizards and sorcerers in their court; that does not make magic real." Pauli argued.

"Perhaps you should try it for yourself." Jung suggested.

Pauli huffed and rolled his eyes.

"The reason it has been around for so long is that it seems to work. Even for people who don't believe in it." Jung added.

"Okay fine, I'll try it." Pauli relented.

Jung pulled out three coins and to Pauli's surprise they were not ancient or Chinese, they were regular coins like the ones Pauli had in his own pocket.

Pauli immediately shook the coins and quickly tossed them up, each flipping in the air before chaotically bouncing to the floor.

"Actually, you skipped a step." Jung said as the last coin settled. "Imagine you are with an oracle. The thing is, neither of you can speak directly to the other. You speak to the oracle by feeling your thoughts and the oracle speaks to you by controlling how the coins land."

"Really?" Pauli asked, unimpressed with such an imaginative divination trick.

"Yes." Jung said patiently as he nurtured Pauli's understanding.

"And how does one feel their thoughts?"

"Think of the question until you feel yourself really wanting the answer," Jung said. "The ancient Chinese knowledge this is based on is an understanding of how reality works. You will find it surprisingly similar to many aspects of quantum physics."

"I doubt that." Pauli remained skeptical.

"They believe there is a flowing energy that creates reality. Could this be the Planck energy flowing through the quantum fields?" Jung challenged.

"Hmpf."

"They call it the Tao and it influences all things." Jung said.

"The Tao," Pauli said, trying to understand, since he truly was searching for an explanation of the Pauli Effect.

"There are two types of Tao energy; yin and yang, positive and negative, male and female, light and dark." Jung explained.

"A balanced, flowing energy." Pauli offered.

"You are going to cast the coins six times and we will record the results which will be yin or yang. This will give you a pattern that has an associated meaning."

"So, I throw the coins then read someone's opinion on what the results mean?" Pauli's nature was to be critical; it was harsh but it produced better scientific results.

"You missed the first step again," Jung instructed. "The Tao is responding to your question to the oracle."

"I just ask it any question?" Pauli thought, "okay, will I make another great discovery in physics?" He said loudly as if he was speaking to some ghost above him.

"To that," interrupted Jung, "the oracle would say, who knows? I cannot tell the future."

"What?!" Pauli was losing his patience with this nonsensical game. "Then how do I ask a question?"

"The oracle is the flowing energy," Jung explained. "The Tao is not the future; it is the energy flowing into the future. I originally pictured myself on a boat in a river. The boat is my life and the flowing river is the Tao. After all our discussions I now understand we are also the same flowing energy, but I digress." [FACT 99]

"Then what kind of question do I ask this flowing energy?" Pauli said.

"It could be for guidance, but it cannot be a yes or no. The final step is for you to interpret your own reading," Jung explained. "This is actually where I got the idea to have clients interpret their own dreams. I can help figure out what symbols mean but only the dreamer can know the true meaning."

"A question for guidance." Pauli mumbled.

"Do you have a question in mind?" Jung asked.

Pauli calmed himself with a deep breath, allowing his defenses to drop. He felt vulnerable as the thought of his failed marriage and his mother's suicide came to mind. He was lost in all aspects of life except his work. His feelings dwelled in his empty personal life and the pain that lingered from past relationships. His mind had only one question and the feeling of his empty bed and missing mother radiated sorrow in his chest; what do I do now? He thought.

"I have my question." He said begrudgingly.

"Then please, hold the idea of your question in your mind and cast the coins." Jung instructed as he readied his pen to record the results.

Pauli tossed the three coins which landed one heads, two tails.

"Seven." Jung said, drawing a horizontal line.

"Seven?" Said Pauli, "there are no numbers here."

"Every heads is worth three and every tails is two. I will explain the rest as we go." Jung instructed. "One more thing, as you cast the coins, really focus on your question; feel the wonder of the question in your heart." He instructed as Pauli tossed the coins five more times.

"Your results are seven, nine, seven, nine, seven, seven." Jung announced as he drew a tower of lines from the bottom up. "This is called a hexagram."

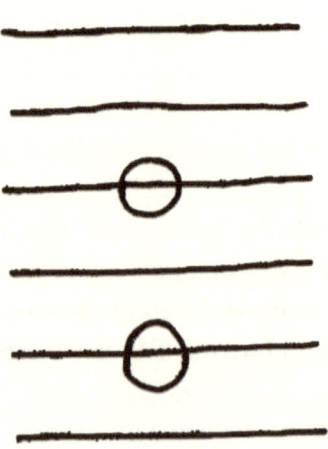

"Why do those lines have circles?" Pauli asked.

"Those are changing lines. It means you have one reading for where your energy is or has been and you have another reading for where your energy is going." Jung explained.

"My energy? You mean my Tao." Pauli clarified.

"Exactly. This is where the book gets its name. The only certainty in life is change and you are about to explore the changing Tao of whatever question you were pondering."

"Okay," Pauli said, trying to allow himself to believe in this sort of nonsense.

"There are 64 possible hexagrams which are made of an upper and a lower half. Your old lines are showing 'heaven' on top and 'heaven' on bottom."

"What does that mean?" Pauli asked as Jung opened a book where he found a table.

"This is hexagram one. It represents your creative energy. We can look up the description in a minute. Let's find the young hexagram first." Jung changed the lines with a circle.

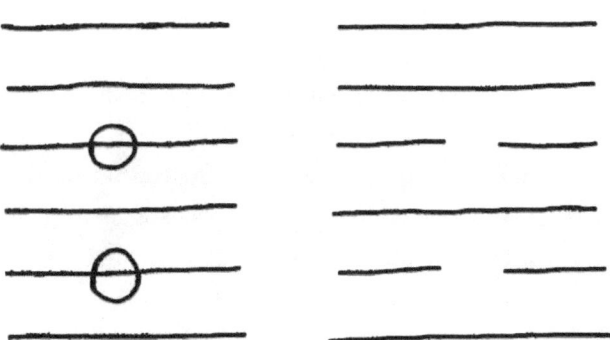

"Wind over fire." Jung said as he ran his finger through a table of hexes to find the intersection. "Here it is. Family, clan, or possibly meaning community; hexagram 37."

365

"What numbers did you just say?" Pauli said in absolute shock. "One hundred thirty-seven?" He said faintly.

"Yes," Jung said calmly as Pauli was overtaken by a rush of adrenalin before an odd feeling of comfort started in his chest and settled deep in his stomach as if something from beyond his understanding was wrapping him in a hug.

Jung wasn't sure why Pauli was so shocked, but he smiled and said, "Does the Tao have your attention now?"

Pauli was without words. The number 137 was a regular fixture in Pauli's life, appearing randomly and even in his work. [FACT 100] This was a synchronicity he could not ignore.

"My subconscious mind did not influence how the coins landed." Pauli reminded himself. "That was controlled only by the laws of nature in objective reality so how would my subconscious mind have sent me this message?" Pauli challenged Jung.

"A synchronicity." Jung smiled. "The internal and external worlds must be reciprocal."

"Meaning?"

"Somehow your unconscious mind or your greater consciousness did influence the coins." Jung said.

"Consciousness influencing external subjective reality." Pauli said calmly, trying to understand.

"We each must have a consciousness that exists external to our physical bodies."

"How? Where?" Pauli said with the frustration setting in. He knew deep in his heart he truly believed what Jung was suggesting but he also understood he would never scientifically prove it.

"I haven't quite figured that out but it seems that with everything you've taught me about quantum physics, it is scientifically possible." Jung said.

"Scientifically possible just means it has not been disproven yet." Pauli said abrasively.

"And yet you know it to be true just as much as I do." Jung said calmly, hoping to comfort the frustration he saw in his friend's eyes.

Pauli opened up to Jung and told him the question he had asked the oracle. The two men read through the descriptions of 'create' and 'family.' After much conversation, Pauli concluded the coins delivered a blatant message; he should create a new family.

After Pauli's counseling session came to a close the men began discussing physics. "I have no more lecture topics for you so do you have any thoughts on what we've already discussed?" Pauli said.

"I am still reconciling my understanding of how everything fits together." Jung began. "Remember that analogy I used of the three different realities?"

"Here we go again." Pauli huffed and rolled his eyes.

"Physical reality would be one reality which all resides on this side of the Higgs mechanism." Jung explained. "The energy in the Planck field, interacting with the quantum fields would be the second reality and it would also be the bridge to the third reality. The final reality would be some higher dimensional quantum-reality, opposite ours that hosts the source-energy and where our spirits originate."

"Spirits!?" Pauli shook his head in disapproval as Jung continued talking.

"Our reality ends at the quantum field; you said that yourself." Jung said.

"No, I didn't," Pauli protested.

"You said virtual particles don't exist in our reality but they are energy in quantum fields which means quantum fields are the limit of our reality!"

"Virtual particles do not exist, not because of their size or location but because of time. They exist for less than one Planck time." Pauli reiterated.

"But time is relative to space," Jung argued. "So could that be the same thing as saying virtual particles exist in higher spatial dimensions?"

"Perhaps there is some value in this awful analogy after all." Pauli said.

Jung was so surprised that he laughed as he leaned back in his chair.

"I still don't like it," Pauli explained, "because it is not some other reality, but I see how it is helping you conceptualize it all. Human Reality is four-dimensional space-time and we do seem to be confined to it since we cannot access anything smaller, or more fundamental, than the energy in the quantum fields. So yes, you could think of it as a reality to itself. This four-dimensional space-time reality exists in a reality with higher dimensions but it is not some *other* reality. We can't access it because only the most minutely small aspects of reality exist in higher dimensions." [FACT 101]

"What if we can access it?" Jung asked slyly.

"What are you getting at?"

"What if consciousness is similar to a force and has a quantum particle generating it? Only that conscious particle is like a virtual particle that somehow exists in higher spatial dimensions? What if we can, in some way, influence or send messages into higher dimensions via our consciousness."

"Sure," Pauli said sarcastically, "and while you're at it why don't you call me a monkey and I'll join the circus! What you are saying is ludicrous, it is in no way scientific!"

"This conversation has transcended the strict confines of science." Jung blurted out. "I cannot hope to prove or in any way test for this, but this is possible! What if our souls are, let's say, 9- or 10-dimensional spirit-energy. Our existence there would be profoundly broader than what we know here in 3D."

"4D if you count time." Pauli added.

"Right. Well, what if the natural laws of the universe only allow physical matter to exist in this three, sorry, four-dimensional space-time reality." Jung suggested.

"You're suggesting the laws of physics prevent any material thing from existing in higher dimensions." Pauli understood.

"Which would leave our physical bodies, and potentially our consciousness, trapped like islands, unable to directly experience anything in the higher dimensions, even if it was part of us."

"Interesting." Pauli said with a curious tone.

"Maybe I missed something," Jung said. "When you originally told me about the four dimensions of space-time, I thought you told me time could not be thought of as a fourth spatial dimension but now you are speaking as if it is."

"Ah, allow me to clarify. We don't know exactly," Pauli said simply. "It is possible that time is a higher spatial dimension. All we know for certain is that space and time are connected to each other in some relational way. We can accurately describe our reality as four-dimensional space-time because we have three spatial dimensions and one temporal dimension."

"Space and time are related as General Relativity suggests." Jung recalled.

"If time is a product of some higher spatial dimension, then perhaps it would help if I explained that."

"Alright," said Jung.

"Remember that two-dimensional napkin world?" Pauli asked.

"Of course." Jung said as Pauli found a piece of paper to represent a two-dimensional reality.

"Well watch this," Pauli said as he slowly moved the paper up. "The paper exists as two dimensions in a three-dimensional reality. If that whole reality was moving upwards, then it would be a two-dimensional reality moving through a three-dimensional reality without being three dimensional itself."

"Oh," Jung exclaimed. "Our whole three-dimensional reality could be moving on a trajectory through a fourth spatial dimension and that is what we experience as linear time." He understood. "Well, which time theory is more likely to be true?"

"Actually, there are problems with both so-"

"We need more information before we realize the truth!" Jung completed Pauli's thought.

Pauli chuckled as Jung's understanding of physics flourished. "Very good," he said.

"I now understand our reality is four dimensions of space-time so to avoid confusion I think it is best to continue calling it 'human-reality.'"

"Why not call it objective reality?" Pauli asked curiously.

"Human-reality is objective reality *and* consciousness or whatever force influences consciousness and the collective." Jung explained.

"So human-reality is objective reality with the assumption that a collective consciousness is manifesting or influencing it." Pauli wasn't sure how he felt about that.

"Back to the three-reality analogy," said Jung.

"Mm hm," huffed Pauli.

"I suppose you've been right about that all along." Jung relented.

"Ha! Yet again I am proven correct." Pauli said victoriously.

"There is only one reality that may have many higher dimensions. It includes everything, all there could ever be." Jung pontificated. "All lower dimensions are physically inside this one great reality; but perhaps 4D human reality is the only place that can produce matter. What I described as another reality is the process itself. The process of higher dimensional energy becoming matter in objective reality."

"Interesting idea." Pauli pondered.

"From our perspective, human-reality seems like one complete reality because we are confined to it," said Jung. "Our consciousness is confined to our physical bodies."

"Like how quantum particles can stabilize, becoming stuck in reality until something else interacts with it." Pauli said without thinking.

"Yes!" Jung exclaimed. "It is exactly like that. No wait, it literally is that! Our bodies are energy that is focused into physical reality by natural processes and laws of physics!"

"And?" Pauli was now curious as it seemed Jung just made some realization.

"Don't you see?" Jung asked. "How did we come to exist in this 4-dimensional space-time human-reality?"

"What are you getting at?" Pauli had no idea.

"Through natural processes! Biology, evolution and physics!" Jung exclaimed. "The natural processes of biology are how our higher dimensional souls exist in human reality. Energy, and therefore matter, cannot be created nor destroyed, only changed!"

"You are suggesting before we are born, we exist as beings of pure energy in a higher spatial dimension but we want to exist physically, so we somehow manifest ourselves into human reality?" Pauli thought about it and it kind of made sense.

"Yes, our souls can only enter this reality through biological processes because matter cannot be created from nothing. I mean why does any life exist at all? Life itself is very counter-intuitive. There must be some process in the quantum fields, that allow us to focus our soul's energy into human-reality, stabilizing our energy as physical matter just like how a quantum particle stabilizes!" Jung culminated.

"That is an intriguing thought," Pauli said. "As you said, energy cannot be created or destroyed so it would take some action in order to focus one's soul into physical reality."

"From our human-reality perspective it would be like a window, a period of time, in which the soul could manifest a physical body." Jung postulated. "Perhaps at birth."

"Or inception." Pauli added.

"Then part of our soul is trapped in this physical existence until we biologically die which releases our consciousness from the confines of physical reality." Jung postulated.

"Because the consciousness loses its mass; a very interesting theory indeed." Pauli said as he allowed the thought to permeate the deepest recesses of his understandings of the natural world. "Since energy, and therefore matter, cannot be created nor destroyed we required a biological process of reproduction to achieve this baggage of a meat-suit that you and I wear around town. Otherwise, we would be like ghosts desperately trying to influence physical reality."

"Or just trying to experience the many sensations of existing," Jung added, guessing that all our physical and psychological sensations can only be experienced in human-reality.

The men pondered the implications of this possibility until Pauli realized he had an engagement he was nearly late for. They said their weekly goodbyes and parted ways.

Instead of walking, Pauli rushed back to the University via streetcar. As the doors opened on both sides of the car he stepped on, turned around and planted his feet in the aisle, holding the railing above. Before the doors closed a pigeon flew through the car, in one door and straight out the other side. Pauli was startled but noticed a man facing him from down the aisle who didn't seem to notice the bird.

"Did you see that?" Pauli demanded.

The man blinked and looked Pauli in the face before saying, "see what?"

"Dammit he was right again." Pauli mumbled as he realized the pigeon flew through the man's line of sight, but he did not notice it. Jung was correct, Pauli thought, we do not consciously notice everything our eyes see. Is there some message for me because I noticed the pigeon? Pauli wondered to himself. No, I saw it but I did not feel any sort of connection to it.

FACTS

97.   Human memory does not work like your computer. Instead of a storage bin that holds memories, they are more like a process. The areas of your brain become active as you remember things just as if you were experiencing or imagining them. This contributes to why our memories are not as good as you think they are. Remember where you were on September 11th, 2001? Your memory is generally correct but the specifics have most likely been tweaked by your memory without you even knowing it. [Ref 77]

98.   Symbliminal is a word created by the author to describe the phenomena when things are seen in objective reality but have a symbolic message to the observer.

99.   If, in some deeper (currently unrealized) recess of reality, the energy-of-creation is the way they describe the Tao, then energy is always flowing, but that does not require each particle's energy to always be flowing. Each quantum field itself requires a minimum amount of energy (one quantum of that particle type) to always be present in the field.

[The previous paragraph is fact (or accepted theory) the following paragraph is the opinion of the author.]

The minimum required energy for each quantum field may be a conglomerate of many discretized energy packets constantly flowing but canceling out, just like how each proton and neutron is many, many fundamental particles constantly canceling out, leaving only the two and one, up and down quarks as its main constituents.

100.   Pauli reported seeing the number 137 constantly throughout his life. It pushed its way into Pauli's conscious mind by way of synchronicities and symbliminal messages. It appeared spontaneously on signs, in papers, or spoken to him, it even showed up in his work. The Fineman Constant works out to 1/137 which Pauli agonized over because (to this day) no one knows why it works out to that specific number. [Ref 79]
    There is no account of Pauli doing the I Ching with Jung, however Jung did use the I Ching as a tool with many patients. How to use the i Ching is accurately depicted here.

101.   The narrative is blending QFT with String Theory. String Theory requires more than 4 dimensions of space-time, usually 9 spatial and 2 temporal dimensions, which can only be accessed by the very smallest bits of our reality. Quantum Field Theory only requires 3 dimensions of space and 1 of time, just as we experience it, however, QFT's math works out to accurately to support higher dimensions. [Ref 73]
    It is the opinion of the author that whatever theory replaces QFT as the next deeper misunderstanding will require reality to be more than 4 space-time dimensions.

Chapter 33

Return of the World Soul

"And once you are awake,
you shall remain awake eternally."

-Friedrich Nietzsche

Another week passed and the usual counseling session closed on a high note.

"Herr Pauli, if you agree, I believe your psychological therapy is complete." Dr. Jung said.

Pauli smiled and took a deep breath before saying, "Yes, I do believe I am ready to live my life. I appreciate the help you have given me grieving my mother and coming to terms with my own actions."

"Unfortunately, I have another client coming in immediately after your session today," Jung said, "but I planned ahead." He pulled out a letter, handing it to Pauli.

"Ha!" Pauli laughed, "I thought I was going to disappoint you today because I have a symposium I must attend at the University. I too prepared a letter! This is a synchronicity if I ever saw one."

"Yes, I would say you and I have unknowingly been planning to be in synch since we started writing these letters!"

"I look forward to reading your letter and will reply when I am able."

"I will do the same." Jung said.

"Auf Wiedersehen Doctor Jung."

"Good travels to you Doctor Pauli!"

After Wolfgang Pauli completed his treatment with Doctor Jung the two rarely met in person but maintained correspondence via letters for the rest of their lives. [FACT 102]

Doctor Jung,

I must admit I am torn by the idea of consciousness. My logical sense tells me that its definition must be limited to our individual waking experiences, but in counseling you proved there was more to it than that. Not only did I have thoughts and feelings from some unknown part of my mind, but I was driven to act in a certain way because of it. This by itself has already displayed an extra-consciousness, though not extra-corpus.

It is a significantly larger claim to say that consciousness somehow influences the manifestation of our reality but that claim now rests with your fundamental questions; what is the consciousness beyond our everyday experiences and what is the fundamental energy that becomes all other energies?

You explained that consciousness very clearly affects objective reality by use of our own free will influencing things that are perfectly within our conscious control, like your mind using its tools (your hand) to pick up the cup of coffee. Then you went on to discuss a case study which may indicate consciousness existing without a physical body.

It is dreadfully unfortunate that your odd cousin was not able to produce any results when the doctors were present. Since we don't fully understand consciousness, if someone proves there is some form of consciousness outside the human body then our entire understanding of reality will be rewritten, this is no small feat. Though I will add that if the spirits were conjured by your mother and cousin then we cannot rule out some part of their physical body was doing the conjuring. Therefore consciousness, even if extra-corpus, may still rely on some aspect of a living body.

I was identified yet again, as the cause of another malfunction. A friend at the University of Göttingen was conducting an experiment and randomly experienced a total failure of his equipment. He later wrote to me saying they could not blame it on me this time. However, as I reviewed my whereabouts, I realized at the exact moment his equipment failed I was stepping off a train, having just arrived in Göttingen! [FACT 103]

For a long time, I felt the Pauli Effect was a curse but after you enlightened me to its implications I now revel in each instance.

"When that amusing 'Pauli effect' of the overturned vase occurred, on the occasion of the founding of the Jung Institute, I had the immediate and vivid impression that I should 'pour out water inside' (to use the symbolic language that I have acquired from you). Then when the connection between psychology and physics took up a relatively large part of your talk, it became even more clear to me what I was to do." [Real Quote from Pauli to Jung]

Many thanks and kind regards.
Yours sincerely, W. Pauli

Herr Pauli,

I do understand Schrödinger's cat was an example of a particle, but I cannot help but wonder what it would be like if a person could experience superposition. I am torn between whether they would simply experience a moment of indecision or would it happen outside of their experience altogether. Is there any math or evidence that prevents the possibility that larger systems could be entangled?

I suspect that consciousness somehow entangles the whole of reality. A literal connection just like how the quantum fields literally connect us all. Thusly allowing for the butterfly effect where great outcomes are influenced by the single swing of a butterfly's wings in some other place long before.

Proton Positivity

You taught me that every type of atom is made of the same three building blocks. One positively charged, one negatively charged and one with no charge at all. The positive one being the only one that determines what type of element the atom is. One proton in an atom makes hydrogen no matter how many neutrons or electrons it has. Two protons in an atom make helium no matter how many neutrons and electrons and so on.

This is perhaps perfect symmetry for a principle of the Law of Attraction. They say that positive thoughts and emotions are significantly more powerful than negative ones (Negative meaning non-existent). They do not provide any scientific observations for why they make this claim, but I cannot help but see the similarity here. In physics only the positive particles determine whether the atom will be gold or brimstone (sulfur that is).

In the Law of Attraction, they say you should only think about the things you do want, in other words the positive. The alternative option being to think of the negative and negating it. For example, instead of saying, "I don't want to get in a car accident." They suggest saying "I want to arrive at my destination safely."

In the former example the thought of a car accident is positively created in the mind. In the latter example, it is only the good image of the real goal being achieved.

It is only the positive particles that control what atomic-reality is and according to the Law of Attraction, the positive thoughts (in a binary sense) have more power to manifest reality. This begs the question...where does the proton's positive charge come from?

I have thrown out my conceptualizations of the three realities and the tree with three branches. They have been replaced with a concept I call Unus Mundus, or One World. It is a single interpretation of everything we have discussed. A combining of the external physical world, described by physics, with the internal world described by psychology.

Have I told you about a phenomenon called the Near Death Experience? There are many accounts in the psychological community and I even had a patient of my own who had this odd experience. In this case he literally died but was resuscitated with a blood transfusion. Later he described this vision, almost like a dream but with even greater clarity than waking life.

He was able to tell me exactly how the scene of his own blood transfusion played out even though, at that moment, he was at least unconscious, but I believe medically dead. If this was the only known case I would say it was just a dream but there are many instances of these experiences from all over the world.

I look forward to hearing from you soonest.
With best wishes,

Yours sincerely,
C. G. Jung

Pauli's Response

Professor Jung,

Please tell me more about this odd phenomenon of the Near Death Experience. It appears you are suggesting there is a phenomenon that could qualify as scientific observations of extra-corpus consciousness. Why have you not spoken of this before? If there is just one proven observation, then consciousness must be more than our everyday experiences and it would require our understanding of existence to be fundamentally rewritten. I have been theorizing some other thing, like a force, which relates to, but fundamentally is not, consciousness. This would change that theory. If anyone has ever died and been resuscitated possessing knowledge of the time during their death, then it is not some other force; it is consciousness.

You have asked an interesting question about a human becoming entangled. Entanglement requires superposition and humans are quite localized. Though, with the idea of Near Death Experiences fresh in my mind, if consciousness exists in a higher spatial dimension, then it could be entangled. In that case it would not be the human that is entangled but the spirit. Of course this is all speculation.

For all those questions I have only one answer, and unfortunately for you, it will cause more questions. Scientists have successfully entangled a living creature. A tiny, microscopic creature called a tardigrade. It was inserted in a series of particles (as if it was a particle) then the whole structure was entangled and the tardigrade lived after the fact. [FACT 104] *What its perspective of this entanglement was, I can only guess that it noticed nothing whatsoever.*

For your proton positivity conundrum, it seems you are confusing the terminology. You seem to be using the term 'positive' to mean good outcome and 'negative' to mean bad outcome but the positive and negative aspects of particles have nothing to do with good or bad.

I believe the origination of the positive and negative symbology was essentially random. In other words, we needed a way to describe the nature of particles and someone came up with positive, negative, and neutral because that conceptualization fit the observed results.

    I know I have a personality that some find abrasive but I am on a quest for truth. I am not concerned for feelings. I have just been informed this has earned me a nickname amongst my peers - Die Geißel Gottes, meaning 'The Scourge of God.' I rather like the epithet.

    In future correspondence please include a better description of Unus Mundus.

    Kind regards.

Yours sincerely,
W. Pauli

Jung's Response

Die Geißel Gottes,

     I applaud your ownership of the name bestowed upon you by your colleagues.

     Proton Positivity. You misunderstood my question. I was not asking about how protons were named, instead I was thinking of the quantum particles that conjure them. Why do two up quarks mixed with one down quark (a proton) determine what element an atom is while one up quark mixed with two down quarks (a neutron) do not?

     Since you brought up how they were named I offer the following. If something with a positive electric charge was randomly assigned the moniker of 'positive' then I must tell you the idea of something positive meaning 'good' was also a randomly assigned moniker. As you said, it just fits.

     The thing is, words and their meanings were not randomly assigned. Over the course of human evolution, we started communicating by making random sounds. We then combined sounds and assigned meaning to them which gave us words. We did not choose random sounds to mean random things. Our ancestors spoke certain sounds and assigned meaning based on how the sound made them feel. [FACT 105] We used our internal conscious world of feelings to express the external objective world via the magic of language.

     Finally, I was not confused on the terminology of positive and negative. Perhaps you should read my first letter again. In my example the idea being thought of was getting in a car accident, which is not what you want but it is the 'positive' because it is the existing thought as opposed to the negative (non-existing) thought.

     When you thought about how to study consciousness you mentioned a vessel that would carry it. Would that be similar to the particles which carry forces? If consciousness is like a force, then there should be some particle to conjure it just as light has the photon. I find this thought very intriguing and would also like to add that instead, consciousness might be a sub-quantum energy or some form of Planck energy.

Perhaps there is some inherent aspect to consciousness which allows it to indefinitely remain as pure pre-existing energy that has influence over other energies in the same way those virtual particles do.

You told me there are laws which are not yet known to us that regulate the quantum fields and the Planck energy. Is it possible there is some law which prevents conscious energy from interacting with the God particle's mechanism? In that case consciousness could never physically exist but it could still influence other creative energies.

I suppose the trouble is if we humans do have some way to connect with that external consciousness it must be through some physical means. We are not missing the particle of consciousness; we are missing an antenna connecting the human body to consciousness in higher dimensions.

Near Death Experiences, or NDEs, seem to be the process of the human death. Each individual has unique experiences, but most NDEs seem to follow a similar pattern. Individuals see and experience different things, but the same elements present themselves in almost all cases. Is that not the definition of a scientific observation? For many years these observations were only from laymen but there have been neuroscientists who experienced and documented these experiences. It is time we acknowledge the evidence of massless consciousness.

Unus Mundus

It is an old concept that captures the essence of my theories, so I have given it new life. It is Latin, meaning 'One World,' first defined by a Catholic friar named Duns Scotus back in the 1200's. I have developed my own *interpretation which now includes the principles of physics you've bestowed upon me.* [FACT 106]

My theory of *Unus Mundus* is supplemental to the quantum field theory which could answer your question; what causes the *Pauli Effect*. *Unus Mundus* is nothing shy of the ultimate process of the creation of the physical universe. Not what created the universe, instead, it's what is creating the universe in every present moment.

I have divided all of reality into two minds; one internal, one external. These minds shape both our physical world and the unseen world of the psyche. In this theory, ultimately everything, originates from a single type of energy. Our consciousness and all of physical reality stem from the same origins that unite the universe in a great oneness.

Here is a simple diagram for a general understanding from our perspective.

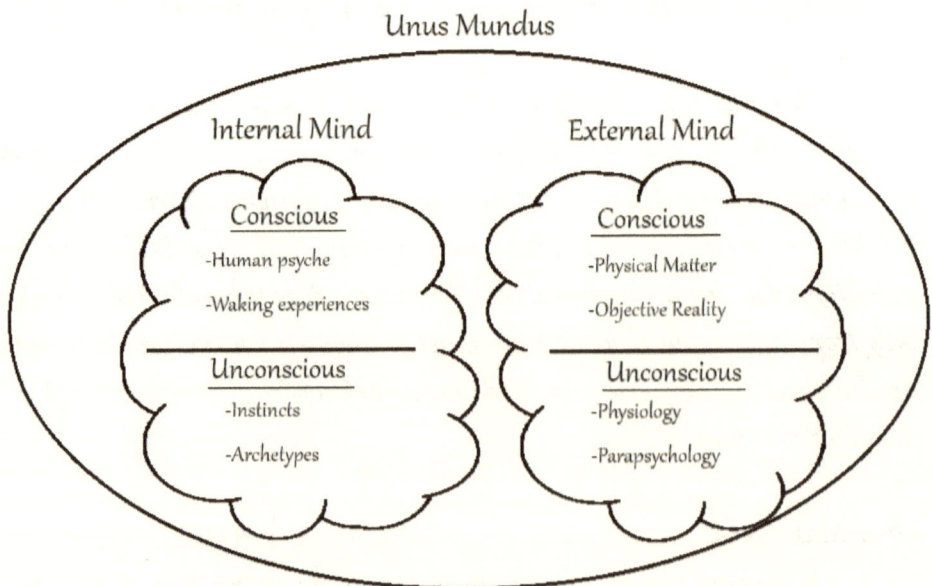

Any one thing in our reality can be put into one of these categories. The internal and external minds both possess conscious and unconscious parts. Looking first at the internal mind, consciousness takes the traditional definition of our own everyday experiences combined with free will. The unconscious part of the internal mind is where we unknowingly possess hidden knowledge, this is our instincts and symbols such as archetypes.

Instincts are an aspect of our internal mind but are physically hosted in our physiology. Physiology is categorized in the external mind's unconscious because your body (brain, DNA, etc.) physically exists as matter in objective reality but we do not consciously experience them.

Parapsychology refers to odd instances of unexplainable things happening in physical reality. A knife exploded in my kitchen, knocking noises echoed from nowhere, and even symbliminal stimuli can be found in the real world as synchronicities. For simplicity's sake I group all this into the category of 'parapsychology.'

Unus Mundus goes much deeper. Both quantum field theory and Unus Mundus postulate that the entirety of reality is literally connected; One World, as the name suggests. I believe there is some greater influence the individual and collective psyche have on influencing reality via this process of Unus Mundus. Though at the moment how to conceptualize it in a way that might lead to experimentation is not apparent to me.

I will now describe Unus Mundus in another way altogether. At its deepest core, Unus Mundus is God. Instead of a man sitting in the clouds, the Christian bible describes God as unfathomable. God is everything and nothing all at once, just as the quantum fields exist without existing. Unus Mundus is the idea of a singularity that is everything that ever was, is, and will be.

Taking a step towards a more tangible definition, there is a collective mind that is somehow an extension of this "God" oneness. The external collective mind is the energy in the quantum fields (and therefore all matter in existence). The internal collective mind is what we see as laws of physics, including the process of creation with the Higgs boson as its precipice.

Perhaps this collective internal mind (the quantum fields' mechanisms) is a membrane that energy passes through as it interacts with the Higgs field to become physically real. This membrane of consciousness would be the "observer" that triggers superposition to breakdown as quantum particles interact with the Higgs field becoming objectively real without the need for a human observer.

It seems to me that the exterior world is directly influenced by the internal world. That means we can consciously influence this creation process via the law of attraction. When conscious influence is not present the universe defaults to unconscious influence. This influence is reciprocated by each individual. In other words, we are all contributing to the collective consciousness that acts as the great observer of the universe's creation process forcing potential to become realized.

Our own existence is due to billions of years of evolution. First the cosmos itself evolved to support biological life then that life evolved to produce us. I think this understanding strengthens my theory of Unus Mundus.

It is us, modern humans, who have been given the gift of free will to take action, even in contradiction to our instincts. We have the ability to use our consciousness like no species has before. As far as we are aware most animals, even today, do not have the ability to think about thinking as we do. It is us who have been given this great gift. Our physiological body is a result of a gradual evolution that has adapted to the laws of physics and biology, giving us both instincts and an abstract consciousness.

It is with this consciousness that we can create a desirable future for ourselves. I would like to point out here, that time is a major conundrum in all this. Understanding what I just described is how we can consciously change our future. Curiously though, we can only experience the eternal present moment.

In conclusion, I will go so far as to claim that when Buddha described achieving nirvana, he was describing a process that combined our individual consciousness with the external collective consciousness which inherently forces the individual to understand they are not an individual at all. Rather we are each the crest of a single wave looking out at all the other wave tops, unaware that we are all part of the same sea.

For my part in it, I believe the collective consciousness is somehow part of, or even is, the quantum field itself. That is Unus Mundus; but unfortunately, the burden of proof is a heavy burden to bear; and, at present, I do not have the means to prove or even test a sliver of it.

We appear to be at a fundamental impasse. We cannot rightly declare that consciousness creates reality because we cannot test it. Likewise, however, we also cannot disprove it. So, anyone who says consciousness does not create our reality is just as scientifically incorrect as someone who says it does. [FACT 107]

With best wishes,
Yours sincerely,   C. G. Jung

Professor Jung,

For your proton positivity, we know that two up quarks mixed with one down quark have a huge influence over our reality, but we have no idea why it is this way.

I've read your original letter and I still feel we are describing two different 'positives.' The law of attraction's 'positive' does not mean good. It means something-rather-than-nothing and the negative simply means, nothing. In electromagnetism the negative is not nothing, it is a force that is opposite the positive force. In your example you described not getting in a car crash as the negative example because of the negative word (not) and because we associate negative to mean bad (but in science there is no good or bad). The way I see it, the law of attraction claims the car crash is positive (meaning something-rather-than-nothing) while the idea of arriving home safely is also positive.

On second thought perhaps it was me who misunderstood. I've just re-read your letter again. I believe I've just proven the point you were making all along. The thought itself, whether of a car crash or of arriving safely is what the law of attraction gurus claim influences reality.

The word 'positive' means good, and something-rather-than-nothing but has also come to describe the proton's charge. These various meanings were all associated with the same word which does seem a synchronous coincidence.

If near death experiences are real, then a whole new field of study should grow to support their exploration. The purpose of that field of study would be to redefine consciousness in a way that incorporates its extra-corpus aspects. This new academic discipline would truly embark on the discovery of the final frontier.

I like your idea that the quantum fields are a membrane and believe you may be on to something with consciousness existing somewhere in the quantum field as pre-energy, unable to interact with the Higgs mechanism because of an undiscovered law of physics.

Finally, I feel we have gotten to the crux of it all. Unus Mundus will need to be refined as the field of study matures.

'Modern microphysics turns the observer, once again, into a little lord of creation in his microcosm, with the ability (at least partially) of freedom of choice and fundamentally uncontrollable effects on that which is being observed. But if

388

these phenomena are dependent on how they are observed, then is it not possible that they are also phenomena (extra-corpus) that depend on who observes them (i.e., on the nature of the psyche of the observer)? And if natural science, in pursuit of the ideal of determinism since Newton, has finally arrived at the stage of the fundamental "perhaps" of the statistical character of natural laws... then should there not be enough room for all those oddities that ultimately rob the distinction between 'physics' and 'psyche' of all its meaning (as with the distinction between 'physics' and 'chemistry')?" [Real Quote, ref 78]

The issue you brought up about defining consciousness is exactly why physicists cannot accept it as a legitimate explanation; we cannot test it. Through my own experiences with synchronicity, I can confirm my suspicions of some greater power or unknown force but it is impossible to define, at least today.

With that being said, we cannot truly prove that gravity exists as a force even though we can measure it (I may have mentioned we have never found the graviton particle) but that does not mean it isn't real.

The burden of proof is quite heavy indeed. Unfortunately for us that is what is required for the world to understand (as you and I now do). However, that is not the only burden here.

In your profession people unknowingly create a 'shadow' for things that will hurt their egos. Bad things they know deep down are their fault (at least partially) but they have come up with reasons for why they are not responsible.

If consciousness really is a factor in creating our reality, then each individual is literally able to influence creation itself, which would require a personal responsibility to many things that people currently take no responsibility for. This burden of responsibility placed on each individual is nothing short of the largest possible burden there could be. If someone living a quiet peaceful life, never had to think past their front garden, was suddenly (partially) responsible for atrocities happening on the other side of the world, that level of responsibility would drive people mad.

This will cause people to reject Unus Mundus altogether, creating an even larger burden of proof on our part. The idea that reality is created by a collective consciousness which we all influence will be denied until it is proven one hundred times over.

Not only are we each responsible for our own experiences but we are literally responsible for the universe and the great process of creation that occurs in every passing moment. Perhaps most individuals are not ready to understand the true power they possess.

I will end this letter with a proposition. As the slow, but thorough, rigors of the scientific method slowly prove what we know to be true, let us now declare this knowledge the 'Pauli-Jung Conjecture' and solidify our place in history.

We postulate that everything in physical reality is being generated from some sort of energy, which we might be tempted to call consciousness but in truth, it needs a new name. Each of us conscious beings can influence and be influenced by this nameless primordial energy. One could describe this theory as a form of pan-psychism, that is, the idea that everything in existence is being generated from the same type of conscious energy. (Though I must add that does not make reality less real, if anything it makes it more special.)

Many thanks and kind regards.
Yours sincerely,
W. Pauli

[FACT 108]

FACTS

102.   Jung would not have entertained conversations regarding physics during Pauli's counseling. The gentlemen depicted in this narrative were from the High Germanic culture of Southern Germany and Switzerland in which everyone remained rigidly within their social queues. It would have been seen as extremely improper for Jung to begin that conversation at all and neither man would have initiated it.

It was in 1934, after completing Pauli's two years of counseling directly with Jung, when the interdisciplinary conversations began and these two renowned greats would push their understandings of their own professions by learning about the others'. The two corresponded mostly by letter. [Ref 80]

103.   The Göttingen Incident. One day, at the University of Göttingen, an expensive piece of equipment was being used in an experiment but stopped working. No cause was ever found and a few hours later it resumed function as normal as if nothing had ever happened. The director of the institute, James Franck, described the incident to Pauli, telling him this time he was innocent. Pauli wrote back explaining that within minutes of that equipment's failure he had been changing trains in Göttingen. [Ref 79]

104.   The tardigrade is a microscopic animal also called a water bear. They have 8 legs and can exist in the most extreme conditions on Earth and even in outer space. In December of 2021 a team of physicists published a paper claiming they quantum-entangled a living organism, the tardigrade, then returned it to normal life conditions where it continued to live with normal functions. [Ref 81]

105.   A word is just sounds. What we are interested in is what our words symbolize. For example, the word 'dog' represents a domesticated creature we often keep as a pet, but you already know what a dog is. The point is the word dog evolved with our relationship with the creature itself.

Each ancient word began as a sound the human mind associated with a certain thing. The next step was communicating the sound from one ancient human to another. Once a shared understanding of the sound's meaning was understood, we had language. Today there are many specific reasons why certain sounds symbolize certain things but the oldest of all reasons was just a feeling. The thing in reality *felt* like the sound they associated with it. [Ref 82]

106.   Unus Mundus.

Jung's idea of Unus Mundus is explained accurately throughout the narrative. At its most fundamental, Unus Mundus is the unknowable source of all reality which we experience through our psyche (internal) and through our interactions with the physical world around us (external). Jung believed all of reality was in one way or another, accessible via our psyche. [Ref 27, 63, 83]

107.   Scientists are actively trying to disprove current theories. This is the very foundation of scientific exploration. The Standard model of particles and QFT are science's most accurate theories and yet scientists are actively trying to prove them wrong by filling in the blanks and expanding their ramifications. Since science acknowledges there are fundamental particles that have not been discovered yet, no one can claim that consciousness is not a factor in manifesting reality. Instead, one could argue (and they do) there are no indicators that suggest consciousness is playing a role, but that claim is simply not true. [Ref 84]

Mainstream science should recognize there are anomalies such as Near Death Experiences that are legitimate evidence of consciousness outside the physical body. Could one of the undiscovered particles be the conscio-boson that carries conscious energy?

108.   Throughout his life, Wolfgang Pauli was generally known as a disagreeable man. Einstein respected him because he knew he could count on Pauli for an honest critique of his work. This willingness to tell people they were wrong is what made Pauli so valuable to Einstein, who had skin think enough to use the criticism for improvement rather than be upset by it. [Ref 79]

Wolfgang Pauli died on December 15, 1958. As his final synchronicity he died in hospital room 137, which had been the number whose meaning eluded him his entire life.

The Feynman constant works out to 1/137 and appears to be the lynch pin that may, one day, unify classic and quantum physics. [Ref 85]

Will you be the one to piece together this quantum puzzle?

Chapter 34

At Both Ends of All Things

"I am the Alpha and the Omega, the First and the Last,
the Beginning and the End."

-Jesus, probably

1944

It was a cold winter morning, typical in Switzerland. Carl Jung, now close to 70 years old, was leaving his house. He walked down his driveway, and just before reaching the bottom his feet lost their grip, giving out beneath him as he slipped on a patch of black ice.

In his next conscious moment, he found himself floating gently through the sky. He was an apparition of pure energy. Floating through the air like a low-hanging cloud on a sunny day. He could feel something calling out for his attention from down on Earth. He saw the hospital where his body was being taken at that moment.

He didn't feel like he was moving but the hospital seemed to be getting further away as if he was slowly walking backwards into a tunnel or a cave. Then, in an instant he appeared at the hospital window, looking into the room where a doctor was standing over his lifeless body. He felt he could have made his body get up whenever he wanted but the gentle calmness of his newfound state-of-reality was too intriguing to jump back into his physical body.

He was invigorated, all the pains he had been accustomed to in old age were gone. He was weightless and free. He started floating higher and higher until he could no longer discern the people or cars below. As he floated through the air, all his senses came together and sights and sounds were experienced with absolutism.

In life, all experiences are filtered through our organs. Sight is tainted by the limitations of our eyes, optic nerves, and brain. In this odd state, however, Carl experienced everything as it truly was. Colors seemed more vibrant and each sound rang out like a note in a wonderous song. He looked into the clouds and fixated on some water vapor. He *became* the water vapor. He could feel the tug of the magnetic pull from other water molecules as they whizzed past him in the cloud. He was one droplet, and he was also the entire cloud.

He felt whole and was curiously excited. He asked himself why he was excited and somehow the answer was understood. He was excited because *he* was about to rain. The whole cloud was abuzz with energy. He separated himself from the cloud and kept rising higher and higher until he could no longer recognize individual buildings.

From here he could see land masses and oceans. Magnificent blue seas surrounding various shades of green continents and tan deserts throughout the lands below. He realized he was so high he was about to leave the Earth's atmosphere. He understood that would be his departure from this incarnation. He felt the draw of a great loving power that told him everything to come would be wonderful. He felt nothing but love radiating from his chest as he accepted his fate with great excitement. He thought about his wife and children but was assured they would be just fine without him.

As he prepared to break through his Earthly bonds, he noticed a temple in an overgrown jungle floating in the sky. The edges of the temple were draped in the type of vines and vegetation you'd find deep in a rainforest. It was like an island floating in space, as if it had been cut out of its original location and somewhere a jungle was left with a hole like a missing puzzle piece.

Like a dream, he was instantly inserted into this new setting. It was not a castle, but a spiritual place. It had odd twisting spires which Jung could not fathom a purpose for. There were decorative pillars and different sets of doors controlling the flow into deeper layers of Jung's spiritual self.

At the very edge, sitting on a short pillar, he saw a Hindu man meditating cross legged, in the lotus position but he did not react as Carl passed. He approached the first set of great doors on the outer edge of the temple. As he reached out, the doors automatically opened revealing an antechamber and the doors to the temple itself.

I am entering my own mind, Carl thought. He slowly entered the courtyard and felt his identity melting away. As he stood in the threshold, he noticed many candles framing the temple's ornate black doors which where intricately carved with geometric patterns. With every passing moment he felt all his worldly hopes, dreams, and worries becoming less and less significant leaving only his accomplishments and memories.

The parts of his individual self that were fading away were being replaced by the most wonderfully fulfilling sense of love and acceptance. It was like his life had been a knight's quest and now that the quest was over his armor was falling off.

He knew that when he opened the great black doors, he would learn all the secrets of reality. However, he would also have to shed the last attributes that made him Carl Jung and he would rejoin the eternal Oneness he spent his life trying to define.

Carl reached out to touch the handle that dangled from a hinge. His fingers inched closer to the knob and as he was just about to open it and walk through, a loud voice startled him causing his hand to recoil.

"You are not leaving! You must go back!"

In an instant, Carl understood without question that whoever was speaking was sacrificing himself so that Carl could return to life.

He turned and saw a King surrounded by a golden wreath of energy. Somehow Carl recognized him as the King of the Greek isle Kos. He also recognized the King as someone else, someone from his life as Carl Jung. I wonder why he has taken another form; even in this ethereal state, I can tell it is him. The words the King spoke were gone but the sound, or perhaps the energy of the message itself, lingered as a reverberation, vibrating in some unknown fabric of reality. The temple, the Hindu man, the jungle, and everything faded away except the vibration of the king's voice. [FACT 109]

Carl slowly opened his eyes to find that he was lying in a bed. His hands instinctively grasped for anything and quickly found the rough bed linens. He realized he was in a cold hospital room. It was the same bed he had seen himself lying in. Was that a dream, he thought. That was no dream; it was a near death experience!

He laid in the bed and looked around. He was miserable. It seemed like all the color had been drained out of the world surrounding him. He was nearly irate that he had returned to the realm of the living. The attending nurse happened to be in the room and tried to keep Carl calm. She informed him that he had a heart attack after he slipped in his driveway and was lucky to be alive. All the pain that once plagued him was back in his old body. In addition to that, there were intense pains shooting from his foot up his leg and his chest and arm were sore from the heart attack.

When the doctor walked in Carl immediately recognized the King of Kos in his present life as a physician. He lashed out at the unsuspecting doctor, partly because he had prevented the glories of the Temple from revealing themselves to him and partly because he knew the doctor was in danger.

Somehow the doctor was sacrificing himself to save Carl. What was done, was done yet Carl wanted to save the doctor, so he tried to warn him but none of what he said made any sense. He could not explain how the doctor would perish nor how he came to understand he would.

"That is no way to treat the man who just saved your life!" Carl's wife, Elizabeth, scolded. With that the doctor left the room.

In a passionate retort Carl called out to explain himself, "Damn it all, he ought to watch his step. He has no right to be so reckless! I want to tell him to take care of himself!"

Carl Jung was Dr. H's last patient. Carl was allowed to sit up for the first time since his heart attack on the same day his doctor became bedridden with illness. Then, on 4/4/44 the doctor died. [FACT 110]
1955

Jung retired to his estate South of Zurich in a quiet village called Küsnacht, or in English, Night's Kiss. The property was on the shore of Lake Zürichee which shimmered behind the house. It was surrounded by gardens, and there was a wall of lush, dark green shrubs that lined the long driveway leading out to the road.

It was a beautiful home with a turreted column above the entryway in the center. It had a pale-yellow façade with green shutters and a steep red roof. The forest-green door was framed by an ornate cement engraving like the mantle of a fireplace. Above the door an inscription read; VOCATVS ATQVE NON VOCATVS DEVS ADERIT meaning "Acknowledged or not, God is present." [FACT 111]

Late one evening, he sat at his desk in the study. This room had dark green shelves filled with books like fruit on a tree in a forest of knowledge. To his left was a large window where he could see the evening moonlight dancing on the calm water of the lake. There he read a letter from one of his longtime colleagues.

The Placebo effect, the friend wrote, has just been proven to be true in a legitimate scientific study. It happens when a patient is tricked into believing they are taking actual medication, when in fact, they are taking a sugar pill that should produce no effects. Here's where the magic happens; the fake medicine actually does whatever the patient was told it does. The medicine itself did nothing. It was the patient's belief that produced the results. They have called it mind over matter. I am very curious as to your thoughts on the subject. [FACT 112]

Jung smiled and thought aloud, "this is the beginning of the end of the world as we knew it." He wrote back to his friend, ending his letter with the following...

In conclusion, this placebo effect will come to be understood as the first scientifically proven aspect of the law of attraction. And so, I believe, at the end of all things is Unus Mundus. Through our deepest feelings, our conscious or unconscious thoughts are each contributing to this unfathomable thing some call God. In past correspondence I referred to the collective conscious to be like God but I believe Unus Mundus is a more accurate term. It is some primordial, pre-physical phase of reality.

I have come to understand there is no such thing as a coincidence. Instead, each coinciding phenomenon should be viewed as an expression of Unus Mundus. The eternal present moment is continuously constituting our whole reality through Unus Mundus like light shining through a series of prisms.

As I slowly decay into the physical hardships of my waning body, I take comfort in understanding that Unus Mundus is also the beginning to all things. The organic resources my body has borrowed for this incarnation as Carl Gustav Jung will be returned to the Earth. Where there is death, there will be new life.

In this way all things, whether tangible like my physical body or abstract like the life my body has lived, each thing has two ends; one where it began and one where it ceases. Beyond both those ends there is, or was, or will be another beginning.

On June 6, 1961, at the age of 85, Carl Jung kissed the night goodbye. As his body and all its mass was left behind, he was free from all the physical limitations of life. Mass is what confines us to this magical existence in human-reality and without it, Jung was free, though some unknown part of him was still connected to it.

He lingered on Earth just a bit. His weightless existence gracefully floated up, crossing through a rainbow. Carl grabbed it, momentarily becoming the rainbow. He felt the subtle differences of each color, like different flavors of gentle sunlight warming his face on a cool morning. The colors felt like music itself, playing some beautiful melody as if the rainbow was an orchestra and each color played a different instrument.

Before he knew it, the rainbow was gone, and he was floating in outer space. He looked out into the darkness admiring the dancing stars that were light years away as they playfully winked and flickered. He was overcome by the celestial beauty of nebulas far beyond Earth, in a seemingly infinite sea of crystal light. In the blink of an eye, he became the Earth. He wasn't above or below it. Not next to, or behind it. Instead, he was within the Earth, and it was simultaneously within him. He realized he was also the moon. This cannot be, he thought. These are so far from each other, how have I grown so large?

Before he could ponder that thought any further, he was floating next to our giant sun. Its radiating plasma splashed as it released massive amounts of energy. Carl Jung then had his last conscious thought.

I should travel faster than light.

He looked toward the vast emptiness of space and flung his spirit, accelerating deep into the void. As he went faster and faster, he felt his memories and worries shedding away. The intangible parts of life like his feelings, memories, and consciousness never had mass because they can't interact with the Higgs field but they are all very real energy. All that energy focused by the quantum fields, tangled up with the energy that creates our physical bodies, expresses itself in a reciprocating dance as one whole person.

He was almost at the speed of light; he passed asteroids, meteoroids and other rogue space debris but the distant stars never seemed to get any closer. For some reason he couldn't go any faster. He struggled, pushing harder and harder to get to the speed of light. His unconscious mind petitioned the universe for more energy to make the final jump to light speed as he was entering the final stages of death. The universe responded by stripping him of the last bit of Earthly connection.

He no longer had mass or even any virtual energy from the intangible aspects of life. His memories and fears were gone and his soul realized it was no longer struggling to maintain its speed. Now, without any effort he reached and surpassed the speed of light becoming one with time itself. At first there was only darkness because light could not keep up.

The veil separating life and death is literally the speed of light. On one side was the death of the body and on the other side was a rebirth of the soul. When we walk into the light at the end of our journey it will effortlessly engulf us in pure love.

For Carl, after he momentarily outran the light, he found himself in a place where there was only light. Not a blinding light he had to shield his eyes from, instead, it was comforting. The warm embrace delivered Jung's soul to the great beyond.

Having now crossed over, everything around him became limitless potential. He saw existence from a fourth spatial dimension before some force quickly pulled him through the fifth, sixth, seventh, and eighth, until he was in his true form, in nine-dimensional existence.

He could see and even visit, any place in space or time. Carl Jung was no longer Carl Jung. His spirit was no longer an individual. He was but a glimmer of virtual energy in a higher dimension. He lingered there for an eternity with no age, no gender, no race or any other individual qualities. He was like a single vibrating string of energy that was in perfect harmony with the entire primordial reality. He was alone and yet he was part of everything that had ever been or would ever be.

Somewhere in that vast eternal existence he decided he should continue his work from Earth, experiencing and explaining the mysteries of human-reality. Without any other planning or preparation, he was emerging from a dark tunnel into a blinding light. He had existed through what felt like epochs outside of time itself, and somehow, he was returning to Earth in what seemed like the blink of an eye. With all the karmic energy from a long strand of lives, a new baby was born. This new person once again had mass, confining her soul to this magical existence in physical reality.
[FACT 113]

FACTS
109.   In early 1944 Carl Jung slipped on ice in his driveway. He fractured his foot and had a heart attack, triggering a near-death experience similar to the depiction in *The Owl and the Osprey*. [Ref 3]

NDEs are being studied by serious scientists. One case study looked at four people on life support who showed no brain activity (besides standard life sustaining patterns). Their families decided to remove life support which is where the research began. Scientists recorded all the brain activity during the process of death, and surprisingly, *after death*. Two participants showed an explosion of brain activity which continued for dozens of minutes without oxygen, defying our current understanding of how brains function. No one has attempted to replicate this study because they fear the implications it might bring.

The brain activity matched known patterns of human thoughts and experiences. It is believed these two people experienced a life review, speaking to someone they recognized, moved through a tunnel towards a light and other experiences commonly reported by NDE survivors. All four participants died so there was no firsthand testimony to corroborate their experiences.

Some scientists interpret these results as a clear indication that NDEs and therefore all consciousness is hosted in the physical brain. They suggest NDEs are the natural process of death and science requires no explanation because the physical brain is capable of hallucinating many wonderous things.

All the experiences in NDEs can be attributed to the physical brain; except one. This is perhaps the smoking gun for consciousness-extra-corpus. Some NDE survivors reported visiting friends and loved ones who were not physically near them at any point during their NDE. Doctors verified they were dead during these time periods, not unconscious, medically dead. The smoking gun is the friends and family who corroborate their almost-dead friend's stories. In other words, the person died, left their body where it was and saw what their friend was doing at that exact moment. All types of details have been corroborated; what their friends were doing, wearing, saying and who they were saying it to.

How did people who were medically dead, see and hear their loved one's conversations when their loved ones were in other places, sometimes far away from the person's physical body? Hardcore mainstreamers will take the easy route and claim all those people are lying. If that's true, then mystery solved, however, most of these people were shaken by their experience and gained nothing.

What if one person is telling the truth?

Individual consciousness must exist without the human body. [Ref 91]

In the study, why did two people show brain activity after death and two did not? Perhaps this is why some people remember NDEs. All people may experience NDEs but perhaps only the ones who show brain activity would remember.

One of the most convincing NDE stories comes from a neurosurgeon. Eben Alexander III was in a medically induced coma when he experienced his NDE. According to Alexander's training, while in a coma the brain has no activity besides basic survival functions such as breathing and maintaining the circulatory system. According to the mainstream theory of consciousness, it is impossible to create memories or have experiences while in a coma.

If consciousness does not require the physical brain, then NDEs can easily be explained. To maintain our current understanding that consciousness only exists in the brain, we must say this neurosurgeon did not experience the things he says he did. The thing is, he is not alone.

Over 10% of heart attack survivors report experiencing a near death experience. Most of these reports come from people with nothing to gain by fabricating a story. The University of Virginia has even created an entire department to study the phenomenon.

For the neurosurgeon, he reported all kinds of experiences and memories that were created while his brain was essentially dead. He has since published multiple books describing this experience including his interpretation of what might be the true depths of consciousness. [Ref 15, 16, 17]

His publications:
   -Proof of Heaven: A Neurosurgeon's Journey into the Afterlife, 2012
   -The Map of Heaven: How Science, Religion, and Ordinary People Are Proving the Afterlife, 2014
   -Living in a Mindful Universe: A Neurosurgeon's Journey into the Heart of Consciousness, 2017

If consciousness exists without being physically real then naysayers are demanding physical evidence of something that cannot physically exist. This demand is not unreasonable, after all, even the four fundamental forces physically exist, propagated by their bosons. Everything science has ever discovered has physical evidence: until now. If consciousness exists outside of physical reality, then we must suspend our use of the scientific method and explore this uncharted territory with both curiosity and skepticism. If we do, then one day we might find aspects of these phenomenon that can be tested properly, complete with a nullifiable hypothesis. Even then it may never be fully proven, requiring scientists to commit the worst blasphemy imaginable: a leap of faith.

Officially, anything in this field is pseudoscience. If being labeled a pseudoscientist makes you sad then you are giving too much power to other people's [incorrect] opinions. In 1953 my father looked at a map in his elementary school classroom and intuitively suggested South America and Africa used to be touching. This casual observation triggered his teacher who scolded him because she *knew* he was wrong. In 1953 continental drift was considered pseudoscience. In the 1960s the idea of plate tectonics was finally accepted as the prevailing scientific theory.

Science and math have never disproven the possibility of life after death; therefore no one can scientifically claim that consciousness requires the physical body. Are you one of the people emotionally clinging to antiquated understandings that omit real observations or are you one of the trendsetters who will be proven right in a few short years? [Ref 18, 88]

110.   Dr. H. (shown to Jung as the King of Kos) treated him after his heart attack and died on 4/4/44. Jung felt very strongly that the doctor had traded his own life for Jung's. [Ref 3]

111.   The Latin inscription above Carl Jung's Door: VOCATVS ATQVE NON VOCATVS DEVS ADERIT. (In later Latin some of those V's could be written as U's becoming; VOCATUS ATQUE NON VOCATUS DEUS ADERIT). The literal words translated individually without context are; call but not call God will be present.

This is a reference to when an army wanted to invade Athens so the leader visited the Oracle of Delphi, asking if they would be victorious. The Oracle's response was something like, 'summoned or not, God will be there.'

Jung's own reasoning for this particular inscription is as follows, "The inscription reminds my patients and myself: Timor Dei Initium Sapiente." Which translates to, 'fear of the Lord is the beginning of wisdom.' From Psalms 111:10. [Ref 86]

112.   The placebo effect was first proven in an experimental study in 1955 and is perhaps the most scientifically proven aspect of the law of attraction. Many modern scientists have logical explanations to prevent themselves from acknowledging the law of attraction and thus committing scientific blasphemy. They continue to deny the true magic that allows for the placebo effect to be true; and yet it is true.

The placebo effect has been proven in many studies; however, it is not affective for all ailments. Asthma, for some reason, does not seem to allow for the placebo effect. Does this mean the whole thing is made up? Don't be silly. Never throw out legitimate data points, *especially* when they do not fit your model. That is where new discoveries hide.

Brain imaging studies have found measurable changes in the neural activity of people experiencing placebo analgesia. They can literally see the belief rewiring the brain. Likewise, the placebo effect has been proven to positively influence blood pressure, heart rate, and pain management. [Ref 87]

113.

June 6, 1961

Carl Gustav Jung met death and kissed the night away, finally entering the light behind the black doors of his temple in the jungle, floating through the sky. He was 85 years old. At the time of his death, he was survived by his four daughters, one son, 19 grandchildren, and 'many' great-grandchildren.

After his near-death experience, Jung authored 13 more publications. After his death his works were compiled and revised into 41 more. His legacy is great, and yet, its depth is still being realized to this day.

Jung spent his life answering questions that most people never even thought to ask. For that, he has pushed the limits of our understanding of ourselves and the entire universe. [Ref 3]

114.   *The Owl and the Osprey* described human-reality down to the quantum fields, theorizing the Planck field and the nature of the energy that creates us all. The science was depicted as accurately as possible while being reduced to as simple concepts as possible. If you want to truly understand physics, you should go through the steps to derive the mathematical equations that postulate the theories. Learning the historical timeline of how theories were

conceptualized also leads to a more intrinsic understanding of the physics but forces you to think less imaginatively. It is the opinion of the author that some form of consciousness will be the next great discovery in physics. Our physical reality is inevitably related to some form of consciousness and one day physics will acknowledge it.

Many people have reported Near Death Experiences throughout human history. Many people have ghost encounters that cannot be explained. Jung's kitchen table and a metal knife inexplicably exploding, seemingly from beyond any physically explainable means. Are coincidences really nothing at all or could consciousness be more than it seems?

In History, Repeating Itself Part II, *Treasure Lies Within*, we will dive deeper into many odd facts and the science that could possibly explain them. Don't forget to look out for the same characters in their reincarnated forms.

Colby K.   *History, Repeating Itself*

References

1. Gerald E. Brown and Chang-Hwan Lee (2006): Hans Bethe and His Physics, World Scientific, ISBN 981-256-610-4, p. 338

2. Chase, Chris; *The eerie broken leg coincidence*; USA Today TouchdownWire found at https://touchdownwire.usatoday.com/2018/11/18/alex-smith-broken-leg-redskins-same-day-joe-theismann-similarities-monday-night-football-video-picture-eerie/

3. Jung, C.G.; Memories, Dreams, Reflections. Published: January 26, 2011, Publisher: Knopf Doubleday Publishing Group

4. Jung: A Modern Master. Pg 19, By Anthony Storr, Published 1973, Publisher HarperCollins Publishers Limited

5. Pruitt, S.; How a Wrong Turn Started World War I; HISTORY, A&E Television Networks, July 17, 2018; found at https://www.history.com/news/how-a-wrong-turn-started-world-war-i

6. Klein, Christopher; 2014, A&E Television Networks; London's World War I Zeppelin Terror found at https://www.history.com/news/londons-world-war-i-zeppelin-terror; Retrieved 2023

7. Rimell, Ray (1989). The Airship VC: the life of Captain William Leefe Robinson. Bourne End: Aston. ISBN 0946627533.

8. Visualization for Pilots, found at www.perfectedflight.com/visualization/

9. Crane, Josh; Angel's Glow: From Civil War Folklore To Winning Science Fair Project; WBUR. August 21, 2020. Retrieved August 18, 2023. From https://www.wbur.org/endlessthread/2020/08/21/angels-glow-science-fair

10. Leatherdale, Duncan; Leefe Robinson: The man who shot down a Baby Killer, 2016, BBC News

11. Anonymous authorial, http://www.worcestershireregiment.com/wr.php?main=inc/vc_w_l_robinson_page1

12. Gilbert, Martin; The First World War: A Complete History, Phoenix, 2008

13. Hallion, Richard P. Rise of the Fighter Aircraft, 1914-1918. Baltimore: The Nautical and Aviation Press, 1984.

14. Foulkes, Imogen (30 May 2016). "Switzerland's forgotten role in saving World War One lives". BBC News. Retrieved at https://www.bbc.com/news/world-europe-36391241 on 21 August 2023

15. Alexander, Eben; *Proof of Heaven: A Neurosurgeon's Journey into the Afterlife*, 2012

16. Alexander, Eben; *The Map of Heaven: How Science, Religion, and Ordinary People Are Proving the Afterlife*, 2014

17. Alexander, Eben and Newell, Karen; *Living in a Mindful Universe: A Neurosurgeon's Journey into the Heart of Consciousness*, 2017

18. https://med.virginia.edu/perceptual-studies/who-we-are/ and subsequent links; retrieved on 21 August 2023. An interesting, related story found at https://www.newsweek.com/i-worked-people-who-came-back-brink-death-1575676 retrieved on 21 August 2023

19. Zimmerman and Howell, History of Blood Transfusion; Annals of Medical History; New Series Volume IV, 1932. Found at

https://www.ncbi.nlm.nih.gov/pmc/articles/PMC7945273/pdf/annmedhist148147-0003.pdf August 2023

20.   Hoerni, Ulrich; Fischer, Thomas; Kaufmann, Bettina, eds. (2019). The Art of C.G. Jung. W. W. Norton & Company.

21.   Cambray, Joe; Synchronicity: An Acausal Connecting Principle; INTERNATIONAL ASSOCIATION FOR ANALYTICAL PSYCHOLOGY; Retrieved from https://iaap.org/jung-analytical-psychology/short-articles-on-analytical-psychology/synchronicity-an-acausal-connecting-principle/ on 22 Aug 2023

22.   Jung, C.G.; Synchronicity: An Acausal Connecting Principle (Bollingen Series XX: The Collected Works of C. G. Jung, Volume 8), 25 Oct. 2010, Princeton Univers. Press; Revised ed. Edition

23.   Evers-Fahey, Karen; Towards a Jungian Theory of the Ego, Taylor & Francis, 2016

24.   Anonymous Author, William Leefe Robinson VC found at https://www.stanmoretouristboard.org.uk/william-leefe-robinson-vc.html retrieved on 22 Aug 2023

25.   Hall, Karen. "Sabina Spielrein." Shalvi/Hyman Encyclopedia of Jewish Women. 31 December 1999. Jewish Women's Archive. (Viewed on January 30, 2023) Found at <https://jwa.org/encyclopedia/article/spielrein-sabina>.

26.   Post-Traumatic Stress Disorder Treatment with Psychedelic Drugs, NYU School of Medicine, found at https://med.nyu.edu/departments-institutes/population-health/divisions-sections-centers/medical-ethics/education/high-school-bioethics-project/learning-scenarios/ptsd-treatment-psychedelics

27.   Jung, C. G.; The Archetypes and The Collective Unconscious. 2nd ed. Collected Works Vol.9 Part 1, Princeton, N.J.: Bollingen, 1981

28.   Miller, 137: Jung, Pauli, and the Pursuit of a Scientific Obsession, 2010, W.W. Norton, London

29.   John Searle (2005). "Consciousness". In Honderich T (ed.). The Oxford companion to philosophy. Oxford University Press. ISBN 978-0-19-926479-7.

30.   Human Brain Function, by Richard Frackowiak and 7 other neuroscientists, page 269 in chapter 16 "The Neural Correlates of Consciousness" (consisting of 32 pages), published 2004

31.   Soter, Steven and Tyson, Neil; COSMIC HORIZONS: ASTRONOMY AT THE CUTTING EDGE, New Press. 2000, American Museum of Natural History, retrieved from https://www.amnh.org/learn-teach/curriculum-collections/cosmic-horizons-book/john-michell-black-holes

32.   Guharay, Deboleena; A Brief History of the Periodic Table; THE AMERICAN SOCIETY FOR BIOCHEMISTRY AND MOLECULAR BIOLOGY; found at https://www.asbmb.org/asbmb-today/science/020721/a-brief-history-of-the-periodic-table

33.   Cambridge Dictionary found at https://dictionary.cambridge.org/dictionary/english/chemistry

34.   Young, Ed; An Immense World, 2022 Random House

35.   Baird, Christopher; "Why are the microwaves in a microwave oven tuned to water?" West Texas A&M University: October 15, 2014 found at

https://www.wtamu.edu/~cbaird/sq/2014/10/15/why-are-the-microwaves-in-a-microwave-oven-tuned-to-water/

36.   You Bend Space-Time!; American Museum of Natural History; found at https://www.amnh.org/exhibitions/einstein/gravity/you-bend-space-time#:~:text=Anything%20with%20mass%E2%80%94including%20your,of%20the%20space%2Dtime%20warp. -> ALTERNATE SOURCE -> DoE Explains... Relativity found at https://www.energy.gov/science/doe-explainsrelativity#:~:text=That%27s%20part%20of%20the%20theory,the%20more%20it%20warps%20things.

37.   Anonymous author; The Four Fundamental Forces, Retrieved from https://universe.nasa.gov/universe/forces/

38.   Whiffen, Peter; A Force to Be Reckoned With, The Cambridge Alumni Magazine, Issue 82

39.   Lindorff, David, Pauli and Jung; The Meeting of Two Great Minds, 2013, Quest Books

40.   Evans, Hilary; SLIders: The Enigma of Streetlight Interference; Anomalist Books, 2010

41.   Enz, Charles; No Time to be Brief: A Scientific Biography of Wolfgang Pauli; Oxford University Press, 2010

42.   Carl Jung Letters Vol II, Pages 108-109

43.   Otap, Lenka; We All Travel Through Spacetime at the Speed of Light Predict, 2019

44.   The speed of light definition provided by Encyclopedia Britannica found at https://www.britannica.com/science/speed-of-light

45.   Ashish; *What Would Happen If You Traveled At The Speed Of Light?* Science ABC, 2023 found at https://www.scienceabc.com/pure-sciences/what-would-happen-if-you-traveled-at-the-speed-of-light.html#:~:text=This%20is%20called%20blueshift.,out%20of%20the%20visible%20spectrum

46.   *Does Time Cause Gravity?* PBS Space Time, PBS Digital Studios Aired 2/24/2021; found at https://www.pbs.org/video/does-time-cause-gravity-00kr0q/

47.   Genetic Science Learning Center. "Epigenetics & Inheritance." Learn.Genetics. July 15, 2013. Accessed August 15, 2023. https://learn.genetics.utah.edu/content/epigenetics/inheritance

48.   Center on the Developing Child, Harvard University; "Gene-Environment Interaction" https://developingchild.harvard.edu/science/deep-dives/gene-environment-interaction/

49.   Bowler, J. "This Startup Made a Functioning 'Invisibility Shield'. Here's How It Works" 05April2022; ScienceAlert; Found at https://www.sciencealert.com/these-people-made-a-functioning-invisibility-shield-all-thanks-to-physics

50.   Robinson and Barron, Epigenetics and the evolution of instincts. Science, 7 Apr 2017, Vol 356, Issue 6333, page 26-27

51.   US Army Colonel Seidule, Ty; Was the Civil War About Slavery? 2015 Pager University Foundation; video at https://www.youtube.com/watch?v=pcy7qV-BGF4

52.　　Gudmestad, Robert; What really started the American Civil War? The Conversation Media Group; 2023 found at https://theconversation.com/what-really-started-the-american-civil-war-205281

53.　　Janney, Caroline; "United Daughters of the Confederacy" Encyclopedia Virginia. Virginia Humanities, 2020; Last updated: 2023, May 18; Found at https://encyclopediavirginia.org/entries/united-daughters-of-the-confederacy/

54.　　Frey, D., & Schulz-Hardt, S. (2001). Confirmation bias in group information seeking and its implications for decision making in administration, business and politics. In F. Butera & G. Mugny (Eds.), Social influence in social reality: Promoting individual and social change (pp. 53–73). Hogrefe & Huber Publishers

55.　　Sherman, Mark; "Does Liberal Truly Mean Open-Minded?", Psychology Today, 2011

56.　　The Civil War Facts National Parks Service, Department of the Interior. Updated 2021. Found at https://www.nps.gov/civilwar/facts.htm

57.　　Creamer, John; Inequalities Persist Despite Decline in Poverty For All Major Race and Hispanic Origin Groups; 2020; Page Last Revised December 9, 2021; Found at https://www.census.gov/library/stories/2020/09/poverty-rates-for-blacks-and-hispanics-reached-historic-lows-in-2019.html

58.　　Güntürkün O. *The Conscious Crow*. Learn Behav. 2021 Mar;49(1):3-4. doi: 10.3758/s13420-021-00466-5. Epub 2021 Feb 17. PMID: 33598801; PMCID: PMC7979642. Found at https://www.ncbi.nlm.nih.gov/pmc/articles/PMC7979642/

59.　　Hill, Napoleon. 2007. Think and Grow Rich. Think and Grow Rich. New York, NY: Jeremy P Tarcher.

60.　　Peierls, Rudolf; *Wolfgang Ernst Pauli, 1900-1958;* Biographical Memoirs of Fellows of the Royal Society; The Royal Society Publishing, 1960

61.　　Chase-Miller, Lisa; *What Do You See?* The Pennsylvania State University © 2023 found at https://sites.psu.edu/pscyh256su15/tag/light-from-above-heuristic/ on 23 Oct 2023

62.　　Glasser NJ, Tung EL, Peek ME. Policing, health care, and institutional racism: Connecting history and heuristics. Health Serv Res. 2021;56(6):1100-1103. doi: 10.1111/1475-6773.13888

63.　　Jung, C.G.; *Man and His Symbols*, (1997) Bantam Doubleday Dell Publishing Group

64.　　Campbell, J.; *The Hero with a Thousand Faces* (3rd ed.) 2012, New World Library

65.　　https://www.fnal.gov/pub/today/archive/archive_2012/today12-05-18_NutshellReadMore.html

66.　　Myrvold, Wayne, Marco Genovese, and Abner Shimony, "Bell's Theorem", The Stanford Encyclopedia of Philosophy (Spring 2024 Edition), Edward N. Zalta & Uri Nodelman (eds.), URL = <https://plato.stanford.edu/archives/spr2024/entries/bell-theorem/>.

67.　　Tang, C.L.; *Fundamentals of Quantum Mechanics For Solid State Electronics and Optics*; Cambridge University Press, 2005

68.　　*How a Detector Works*, found at https://home.cern/science/experiments/how-detector-works found October 2023

69.   Lewis, P.J.; *Interpretations of Quantum Mechanics*, found at https://iep.utm.edu/int-qm/ found October 2023 © Copyright Internet Encyclopedia of Philosophy and its Authors

70.   Cooke, M.; DOE Explains...the Higgs Boson; US Department of Energy, Office of Science; found in October 2023 at https://www.energy.gov/science/doe-explainsthe-higgs-boson

71.   LaViolette, P. A.; *The Cosmic Ether: Introduction to Subquantum Kinetics*, Physics Procedia Volume 38, 2012, Pages 326-349

72.   Kuhlmann, Meinard; Physicists Debate Whether the World Is Made of Particles or Fields--or Something Else Entirely; on August 1, 2013 found at https://www.scientificamerican.com/article/physicists-debate-whether-world-made-of-particles-fields-or-something-else/ in October 2023.

73.   Aharony, Ofer; Quantum field theories in higher dimensions; Weizmann Institute of Science, Department of Particle Physics and Astrophysics; found at https://www.weizmann.ac.il/particle/Aharony/research-activities/quantum-field-theories-higher-dimensions-0 in October 2023

74.   Dent, E; The observation, inquiry, and measurement challenges surfaced by complexity theory; Managing Organizational Complexity: Philosophy, Theory and Application, pp.253-268 (2005) Information Age Publishing, 99383431962606570

75.   C.G. Jung Institute History found at https://junginstitut.ch/en/About-Us/History in October 2023

76.   Pauli Effect; Encyclopedia, Science News & Research Reviews; found at https://academic-accelerator.com/encyclopedia/pauli-effect in October 2023

77.   Pappas, Stephanie; *Do You Really Remember Where You Were on 9/11?* 2011, LiveScinece, found at https://www.livescience.com/15914-flashbulb-memory-september-11.html

78.   Jung, Pauli, et al.; Atom and Archetype – The Pauli/Jung Letters, 1932-1958, Princeton University Press; Revised ed. Edition, 2014

79.   Wolfgang Pauli and Carl Gustav Jung, Eidgenössische Technische Hochschule Zürich, found at https://library.ethz.ch/en/locations-and-media/platforms/virtual-exhibitions/wolfgang-pauli-and-modern-physics/wolfgang-pauli-and-carl-gustav-jung.html in October 2023

80.   Fuge, Lauren; *Did Scientists Really Quantum Entangle Tardigrades?* Jan 5th 2022, found at https://cosmosmagazine.com/science/physics/did-scientists-really-quantum-entangle-tardigrades/ in October 2023

81.   Reilly, J., Biun, D., Cowles, W. et al. *Where did Words Come from? A Linking Theory of Sound Symbolism and Natural Language Evolution*. Nat Prec (2008). https://doi.org/10.1038/npre.2008.2369.1

82.   Jung, C. G.; From "The Conjunction", Mysterium Coniunctionis, Collected Works, XIV, New Jersey: Princeton University Press.

83.   https://www.nationalgeographic.co.uk/science-and-technology/2021/04/ultra-precise-experiment-finds-hints-of-unseen-particles-in-the-universe -this explains understanding that the standard model of particle physics is actively being proven wrong – alluding to the fact that there are undiscovered particles out there

84.   Anonymous author, The Mysterious 137 found at
http://www.feynman.com/science/the-mysterious-137/

85.   Jung, C.G. (1975) Letters: 1951-1961, ed. G. Adler, A. Jaffe, and R.F.C. Hull,
Princeton, NJ: Princeton University Press, vol. 2.

86.   The Power of the Placebo Effect, Dec 13, 2021 Harvard Health Publishing,
Harvard Medical School; found at https://www.health.harvard.edu/mental-health/the-
power-of-the-placebo-effect in October 2023

87.   Moody, R. A. (1975). Life After Life. HarperOne.

88.   Watanabe, Satosi (1955). "Symmetry of physical laws. Part III. Prediction and
retrodiction". Reviews of Modern Physics. 27 (2): 179–186.

89.   Wharton, William R. (1998-10-28). "Backward Causation and the EPR Paradox"

90.   Ferreira, Becky; A Growing Number of Scientists Are Convinced the Future
Influences the Past, March 16, 2023; found at https://www.vice.com/en/article/a-
growing-number-of-scientists-are-convinced-the-future-influences-the-past/ pulled
March 2025

91.   Xu, G, Et al, Surge of Neurophysiological Coupling and Connectivity of
Gamma Oscillations in the Dying Human Brain, Proceedings of the National Academy of
Sciences, May 2023 Edited byTononi 2022 found at
https://doi.org/10.1073/pnas.221626812

92.   de Barros, J.A., Oas, G. Can We Falsify the Consciousness-Causes-Collapse
Hypothesis in Quantum Mechanics?. Found Phys 47, 1294–1308 (2017). Found at
https://doi.org/10.1007/s10701-017-0110-7

93.   Smolin, L. and Verde, C; The Quantum Mechanics of the Present, April 21,
2021, found at https://arxiv.org/pdf/2104.09945 April 2025

94.   Dash, Mike; Curses! Archduke Franz Ferdinand and His Astounding Death Car;
Smithsonian Magazine, April 22, 2013; found at
https://www.smithsonianmag.com/history/curses-archduke-franz-ferdinand-and-his-
astounding-death-car-
27381052/#:~:text=It%20can%20be%20taken%20to,without%20spotting%20the%20pl
ate's%20significance.